RECKONING WITH THE U.S. ROLE IN
Global Ocean Plastic Waste

T0201033

Committee on the United States Contributions to
Global Ocean Plastic Waste

Ocean Studies Board

Division on Earth and Life Studies

A Consensus Study Report of

The National Academies of
SCIENCES · ENGINEERING · MEDICINE

THE NATIONAL ACADEMIES PRESS
Washington, D.C.
www.nap.edu

THE NATIONAL ACADEMIES PRESS   500 Fifth Street, NW   Washington, DC 20001

This study was supported by the National Oceanic and Atmospheric Administration under Award Number WC133R17CQ0031/1305M320FNRMA0082. Any opinions, findings, conclusions, or recommendations expressed in this publication do not necessarily reflect the views of any organization or agency that provided support for the project.

International Standard Book Number-13:   978-0-309-45885-6
International Standard Book Number-10:   0-309-45885-4
Digital Object Identifier: https://doi.org/10.17226/26132
Library of Congress Control Number: 2022933130

Suggested citation: The National Academies of Sciences, Engineering, and Medicine. 2022. *Reckoning with the U.S. Role in Global Ocean Plastic Waste*. Washington, DC: The National Academies Press. https://doi.org/10.17226/26132.

The National Academies of
SCIENCES · ENGINEERING · MEDICINE

The **National Academy of Sciences** was established in 1863 by an Act of Congress, signed by President Lincoln, as a private, nongovernmental institution to advise the nation on issues related to science and technology. Members are elected by their peers for outstanding contributions to research. Dr. Marcia McNutt is president.

The **National Academy of Engineering** was established in 1964 under the charter of the National Academy of Sciences to bring the practices of engineering to advising the nation. Members are elected by their peers for extraordinary contributions to engineering. Dr. John L. Anderson is president.

The **National Academy of Medicine** (formerly the Institute of Medicine) was established in 1970 under the charter of the National Academy of Sciences to advise the nation on medical and health issues. Members are elected by their peers for distinguished contributions to medicine and health. Dr. Victor J. Dzau is president.

The three Academies work together as the **National Academies of Sciences, Engineering, and Medicine** to provide independent, objective analysis and advice to the nation and conduct other activities to solve complex problems and inform public policy decisions. The National Academies also encourage education and research, recognize outstanding contributions to knowledge, and increase public understanding in matters of science, engineering, and medicine.

Learn more about the National Academies of Sciences, Engineering, and Medicine at **www.nationalacademies.org**.

The National Academies of
SCIENCES · ENGINEERING · MEDICINE

Consensus Study Reports published by the National Academies of Sciences, Engineering, and Medicine document the evidence-based consensus on the study's statement of task by an authoring committee of experts. Reports typically include findings, conclusions, and recommendations based on information gathered by the committee and the committee's deliberations. Each report has been subjected to a rigorous and independent peer-review process and it represents the position of the National Academies on the statement of task.

Proceedings published by the National Academies of Sciences, Engineering, and Medicine chronicle the presentations and discussions at a workshop, symposium, or other event convened by the National Academies. The statements and opinions contained in proceedings are those of the participants and are not endorsed by other participants, the planning committee, or the National Academies.

For information about other products and activities of the National Academies, please visit www.nationalacademies.org/about/whatwedo.

OCEAN STUDIES BOARD

Preface

The success of the 20th century miracle invention of plastics has also produced a global-scale deluge of plastic waste seemingly everywhere we look. The visibility of global ocean plastic waste, paired with increasing documentation of its ubiquity, devastating impacts on ocean health and marine wildlife, and transport through the food web, has brought widespread public awareness. Recent global attention has made it clear that the ocean plastic waste problem is linked inextricably to the increasing production of plastics and how we use and treat plastic products and waste from their beginning to well beyond the end of their useful lives.

In the United States, ocean plastic waste has become a top public concern, but the developing plastic waste crisis has been building for decades. While U.S. landmark environmental protection laws were enacted in the 1970s to address hazardous waste and toxic water and air pollution, they did not target more widespread plastic waste. Instead, U.S. attention to ocean waste understandably focused on reining in ship- and marine-based sources of ocean pollution, and on controlling discharges of toxic chemicals such as dichlorodiphenyltrichloroethane (DDT) and other harmful and hazardous releases to U.S. air and waters.

Coastal states and remote islands, and those who make their living from the sea, raised early alarms about ocean plastic waste, often referred to as "marine debris." Attention centered on the contributions from lost or abandoned fishing gear and ship-based disposal of plastics and other waste. These calls for action resulted in early government and nongovernmental programs targeting identification and cleanup of fishing gear and other trash on beaches and those harming marine habitats and entangling wildlife. Important land-based sources of plastic waste—a growing proportion of marine debris—were governed at the state level largely under solid waste management controls such as landfills, recycling, or incineration.

After a decade of largely regional efforts to address marine debris, in 2004, the congressionally chartered U.S. Commission on Ocean Policy identified marine debris as a national ocean priority and called for strengthening marine debris efforts at the National Oceanic and Atmospheric Administration (NOAA) and other federal agencies. These recommendations shaped the 2006 Marine Debris Act, which has been reauthorized and updated three times—most recently last year, by the Save Our Seas 2.0 Act. Other laws enacted over time include the 2015 federal ban on the use of microbeads in certain products. Federal attention to land-based sources of ocean plastic waste was constrained in light of other priorities. As a result, ocean plastic waste has overwhelmed current marine debris control efforts, despite the important work all parties have achieved to date.

Since 2000, U.S. federal programs focusing on marine debris and waste management have been gaining attention in Congress. State and local action on ocean plastic waste has been outpacing federal action, with many state and local bans or restrictions on sale or use of plastic items seen most frequently in communities and coastal environments. An accumulating number of scientific studies and expert reports have raised the level of attention to the problem of plastic waste, generally, and ocean plastic waste, specifically.

Global attention to ocean plastic waste accelerated in 2016 when the United Nations adopted a new ocean-focused Sustainability Goal 14 (Life Below Water), which identified the need to address ocean plastic and other sources of ocean pollution. The United Nations Environmental Assembly has passed four resolutions since 2014, including a call for stronger coordination and a shared vision to tackle marine plastic waste. Plastic waste is on the agendas for the G7 and G20, the United Nations, and other bodies, with growing interest in a global treaty on plastic pollution. Many nations are already developing aggressive goals, strategies, and laws to stem the tide. In 2021, the UN marked the beginning of its Decade of Ocean Science for Sustainable Development, with at least one focus area on the problem of ocean plastic pollution. Additionally, international law has been amended to control exports of plastic waste under the Basel Convention on the Control of Transboundary Movements of Hazardous Wastes and Their Disposal, but the United States is not a signatory.

Against this backdrop, the National Academies of Sciences, Engineering, and Medicine (the National Academies) engaged in efforts to understand the issues through consensus studies, including Clean Ships, Clean Ports, Clean Oceans: Controlling Garbage and Plastic Wastes at Sea (National Research Council 1995) and Tackling Marine Debris in the 21st Century (National Research Council 2009). Several years ago, the

Ocean Studies Board (OSB) identified ocean plastic waste as an area of rapidly evolving scientific discovery and societal relevance, and selected the topic for the 21st Annual Revelle Lecture in March 2020, which was delivered by Chelsea Rochman, one of a rising generation of scientists working on the problem. That same month, just before the COVID-19 pandemic reduced travel, OSB held a workshop on the ocean plastic problem, at about the same time that two other National Academies workshops were held on other plastic-related topics: *Closing the Loop on the Plastics Dilemma* (NASEM 2020) and Emerging Technologies to Advance Research and Decisions on the Environmental Health Effects of Microplastics.

In June 2020, NOAA engaged OSB and sponsored this study, grounded in one outlined by Congress in the Save Our Seas 2.0 bill (enacted later in 2020). OSB convened this ad hoc consensus Committee on the United States Contribution to Global Ocean Plastic Waste around an ambitious statement of task. Despite the many challenges of operating during a global pandemic, the committee met frequently to understand the state of knowledge about ocean plastic waste. We focused on specific issues facing the United States, as well as on what solutions are being tested at the local to global levels. The committee benefited from insights from federal programs at NOAA and the U.S. Environmental Protection Agency, and a range of experts and practitioners, as well as U.S. plastic waste priorities and activities. These include the United States Federal Strategy for Addressing the Global Issue of Marine Litter and priorities identified in the 2018 National Science and Technology Council Decadal Ocean Science and Technology Vision, which included preventing and reducing plastic pollution.

Much of the information on plastic waste that the committee relied on came from available government and industry data and a substantial number of studies conceived and carried out by scientists and other experts in nongovernmental organizations and academia, with limited federal support. A hallmark of these studies has been their grounding in collaborations, in partnership or coordination with government, communities, and industry groups. Philanthropic support and insights have injected innovative "circular economy" principles to these collaborations, which may help unite action toward economically beneficial solutions. Community science has grown in popularity, especially among young people. The rising generation is deeply engaged and motivated to raise their concerns about ocean plastic waste to decision makers.

While this report identifies knowledge gaps, it also summarizes what we learned, and lays out opportunities for the United States to stake out a leadership position and take meaningful steps in the United States and on the global stage, with many co-benefits for U.S. policy priorities, from climate change and social equity to economic opportunities and technology innovation. Strategies and roadmaps developed by U.S. states and other nations serve as illustrative examples.

The problems caused in the ocean and for society by the rise of plastic waste are complex and accelerating. Solving them requires a systemic and systematic approach unified around clear goals and paths for change. Ocean plastic waste is part of an overall challenge from the global growth of plastic production, especially based on fossil sources, and related economic trends, along with gaps in waste management. The disparate impacts on people and communities makes equity important in formulating strategies and evaluating impacts, costs, and solutions. The increase in plastic waste with the COVID-19 pandemic underscores the influence of larger global challenges.

As the U.S. public learns more about the plastic problem, it seeks clarity on top causes and key solutions now and for the future. Public outcry and attention in the United States and globally will intensify as more studies and reports are released by scientists and other experts. Public concern has led Congress to call for several studies to delve more deeply into questions beyond the committee's charge. In October 2021, the United Nations Environment Program released a comprehensive global assessment of marine litter and plastic pollution to inform discussions on national and global action on plastic pollution, including a global plastic treaty (UNEP 2021a). These insights will join the growing wave of information and add to our national knowledge base.

This report is a first-order synthesis of what we learned about the questions raised in the statement of task. It by no means addresses all questions or provides all answers, but it does provide some sample blueprints for action. The report provides suggestions for a U.S. plan of action and federal leadership on this problem, including on the global stage. This will require strong federal coordination that draws on the advice and knowledge of a range of experts and practitioners, including those with a deep understanding of the incentives, processes, and practices that must change if we are to prevent plastics from entering our environment and our ocean as uncontrolled and harmful plastic waste.

The committee members and I would like to thank NOAA and the congressional sponsors for their longstanding commitment to addressing the problem of ocean plastic waste. We were honored to be selected for this important task, and I am grateful to my fellow members for their generous contributions of expertise and time. I know they join me in appreciating the tireless work of our study director, Dr. Megan May, and the larger National Academies team. I also thank the members of the Ocean Studies Board and board director, Dr. Susan Roberts, for their commitment to this important topic.

Margaret Spring, *Chair*
Committee on the United States Contribution to Global Ocean Plastic Waste

Acknowledgments

The committee would like to thank National Oceanic and Atmospheric Administration (NOAA) staff and contractors who helped with this project, especially Amy V. Uhrin, MaryLee Haughwout, Nancy Wallace, Emma Tonge, Ya'el Seid-Green, Hannah Montoya, Patricia McBride Finneran, and Ryan Edwards. At five open-session meetings, the committee heard presentations from a wide array of experts. The committee thanks all the speakers: Amy V. Uhrin (NOAA), Senator Sheldon Whitehouse (U.S. Senate), Mary-Eileen Manning (Office of Senator Sullivan), Jill Hamilton (Office of Senator Whitehouse), Stewart Harris (American Chemistry Council), Steve Alexander (Association of Plastic Recyclers), Nicholas Mallos (Ocean Conservancy), Winnie Lau (Pew Charitable Trusts), Scott Fulton (Environmental Law Institute), Mary Ellen Ternes (Earth & Water Law, LLC), David Biderman (Solid Waste Association of North America), Jonathan Bishop (California State Water Resources Control Board), Jeremy Conkle (Texas A&M University Corpus Christi), Timothy Hoellein (Loyola University Chicago), Sebastian Primpke (Alfred Wegener Institute), Shungu Garaba (Carl von Ossietzky Universität Oldenburg), Victor Martinez-Vicente (Plymouth Marine Laboratory), Ellen Ramirez (NOAA), Hillary Burgess (NOAA), Romell Nandi (U.S. Environmental Protection Agency), Harry Allen (Environmental Protection Agency Region 9), Nancy Wallace (NOAA), and Rusty Holleman (University of California, Davis).

The committee appreciates the American Chemistry Council's willingness to share data on plastic production, which were used for Chapter 2.

Finally, the committee would like to thank The Research Center of the National Academies of Sciences, Engineering, and Medicine for its research support, Eric Edkin (the National Academies) and International Mapping for graphics support, and Rona Briere for her editing support.

Reviewers

This Consensus Study Report was reviewed in draft form by individuals chosen for their diverse perspectives and technical expertise. The purpose of this independent review is to provide candid and critical comments that will assist the National Academies of Sciences, Engineering, and Medicine in making each published report as sound as possible and to ensure that it meets the institutional standards for quality, objectivity, evidence, and responsiveness to the study charge. The review comments and draft manuscript remain confidential to protect the integrity of the deliberative process.

We thank the following individuals for their review of this report:

Jennifer Adibi, University of Pittsburgh
Stefano Aliani, Institute of Marine Sciences National Research Council
Eric Beckman, University of Pittsburgh
Winnie Lau, Pew Charitable Trusts
Diane Sicotte, Drexel University
Mark Spalding, Ocean Foundation
Anastasios Xepapadeas (NAS), Athens University of Economics and Business

Although the reviewers listed above provided many constructive comments and suggestions, they were not asked to endorse the conclusions or recommendations of this report nor did they see the final draft before its release. The review of this report was overseen by **Danny Reible** (NAE), Texas Tech University, as the Report Review Committee Monitor and **Michael Kavanaugh** (NAE), Geosyntec Consultants, as the Division of Earth and Life Sciences Coordinator. They were responsible for making certain that an independent examination of this report was carried out in accordance with the standards of the National Academies and that all review comments were carefully considered. Responsibility for the final content rests entirely with the authoring committee and the National Academies.

Contents

Summary

Global ocean plastic waste originates from materials introduced in the 20th century to deliver wide-ranging benefits at low cost. Plastics increased an era of disposability for products and packaging used for a short time and then thrown away. The result has been a dramatic rise in plastic waste, which in turn leaks to the environment, including the ocean. Plastic waste has a range of adverse impacts, some of which are only beginning to be recognized and understood. Over the past decade, research on ocean plastic pollution has revealed that plastic waste is present in almost every marine habitat, from the ocean surface to deep-sea sediments to the ocean's vast mid-water region, as well as the Laurentian Great Lakes. An estimated 8 million metric tons (MMT) of plastic waste enter the world's ocean each year—the equivalent of dumping a garbage truck of plastic waste into the ocean every minute. If current practices continue, the amount of plastics discharged into the ocean could reach up to 53 MMT per year by 2030, roughly half of the total weight of fish caught from the ocean annually.

Society is grappling with the massive scale of the challenge of plastic waste with responses ranging from beach cleanups and local bans to extended producer responsibility schemes, circular economy commitments, country-level plans and commitments, and a call for a global treaty. Decision makers are calling for reliable syntheses of the state of scientific knowledge at national and global levels. This report is designed to provide that synthesis for U.S. decision makers.

The contribution arose from the Save Our Seas 2.0 Act, sponsored by a bipartisan group of 19 senators, which passed into law on December 18, 2020, in the 116th Congress. Among a variety of components in the law, it called for the National Academies of Sciences, Engineering, and Medicine to lead a study examining the United States' contribution to global ocean plastic waste.

The task for the committee was to review data on the size of U.S. contribution to plastic waste generation, waste mismanagement, the paths these wastes take to the ocean, and the distribution and fate of these wastes once they leak into the ocean. The committee assessed the potential value of a national marine debris tracking and monitoring system and how such a system might be designed and implemented. Finally, the committee identified knowledge gaps and recommended potential means to reduce U.S. contributions to global ocean plastic waste.

U.S. PRODUCTION AND GLOBAL TRADE

Over a 50-year period, global plastic production increased nearly 20-fold, from 20 MMT in 1966 to 381 MMT in 2015. The U.S. contribution to global ocean plastic waste begins with the plastics produced and used in this country or exported to other nations, as well as plastics manufactured elsewhere that enter the U.S. waste stream through trade. Petrochemical plants convert fossil-based feedstocks (e.g., crude oil, natural gas liquids) into polymers, while biobased plastics are plastics in which the carbon originates, in whole or in part, from renewable biomass feedstock such as sugar cane, canola, and corn. More than 99% of the plastic resin produced globally is made from fossil-based feedstocks. The majority of plastics are hydrocarbon plastics (from fossil-based or biobased feedstocks). Hydrocarbon plastics have a strong carbon-carbon bond, making them resistant to biodegradation.

Plastics are a family of synthetic polymers composed of resins that have different chemical and physical structures; examples include polyethylene, polyethylene terephthalate (PET), and polypropylene. In 2019, a total of 70 MMT of plastic resin was produced in North America, which can be compared to global production of 368 MMT, according to Plastics Europe. Data for resin production are not available for the United States alone. While trends for different types of plastic resin vary, the overall trend for resin supply and production has increased over the past 10 years. Using the American Chemistry Council data, the committee estimated in 2020 that for eight resins, a total of 41.1 MMT was produced in North America. This estimate is not complete and does not include all plastics produced because the committee was unable to identify data for PET, thermoset, and resin fibers.

In addition to producing plastic resin, the United States imports and exports plastic products. The U.S. trend of both plastic exports and imports has been increasing over the past three decades. According to the U.S. Census Trade data, in 2020 the United States exported 2,342,368 categories of plastic products, defined as "the number of individual export line items," at a value of $60.2 billion. In 2020, the United States imported 5,747,472 categories of plastic products ("the number of individual import line items") at a value of $58.9 billion.

Conclusion 1: Because the vast majority of plastics are carbon-carbon backbone polymers and have strong resistance to biodegradation, plastics accumulate in natural environments, including the ocean, as pervasive and persistent environmental contaminants.

PLASTIC WASTE AND ITS MANAGEMENT

From 1950 through 2017, the world cumulatively produced 8.3 billion metric tons (BMT) of plastics for use. By 2015, 6.3 BMT of plastics had become waste. Annually, the world generates 2.01 BMT of waste, of which 242 MMT is estimated to be plastic waste.

U.S. Plastic Waste Generation

The U.S. per person municipal solid waste (MSW) generation rate ranges from 2.04–2.72 kg/day (4.5–6 lb/day), depending on the reference examined. This is 2–8 times the waste generation rates of many countries around the world. While only 4.3% of the world's population lives in the United States, the nation was the top generator of plastic waste and total waste in 2016, with a total plastic waste at 42 MMT and a per capita plastic waste generation of 130 kg/year (Law et al. 2020).

MSW plastic waste generation has been increasing in the United States since 1960, with the fastest increase seen from 1980 to 2000 (Figure S.1). The steep increase in plastic production has been mirrored by an increase in the percent of U.S. plastic solid waste (by mass)—from 0.4% in 1960 to the 12.2% observed in 2018, with a peak of 13.2% in 2017. While both recycling and combustion as plastic waste management techniques expanded in the 1980s and 1990s, the amount of plastic waste managed using these techniques has not expanded relative to the increase in plastic waste, resulting in more plastic waste in landfills (Figure S.2). Designing plastics at the end of onset so that they can be appropriately managed at their end of life could address this trend.

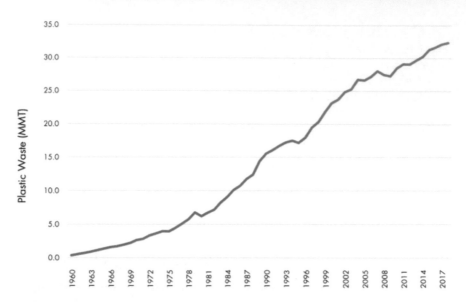

FIGURE S.1 U.S. annual plastic solid waste generation from 1960 to 2018 in million metric tons. SOURCE: U.S. EPA (2020a).

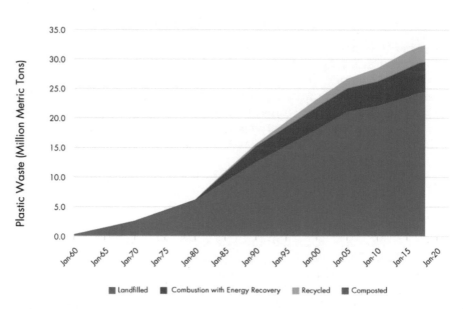

FIGURE S.2 U.S. plastic waste management of municipal solid waste from 1960 to 2018 in million metric tons (MMT) per year. Composted levels are zero during this period. SOURCE: U.S. EPA (2020a).

U.S. Contribution to Plastic Waste Leakage

Solid waste management systems are important to understand the difference between managed and mismanaged solid waste. In theory, solid managed waste should not contribute to ocean plastic waste because it is contained by treatment and/or conversion into other products (recycling, composting, incineration) or contained in an engineered landfill environment. In practice, plastic waste still "leaks" from managed systems through blowing out of trash cans, trucks, and other managed scenarios. In addition, waste not put into the solid waste or another management system, whether intentionally or unintentionally, through actions such as illegal dumping, littering, or unregulated disposal or discharge, also "leaks" into the environment. The U.S. Environmental Protection Agency does not monitor or report on any of these sources of leaked plastic waste.

Even with an advanced solid waste management system, U.S. plastic waste is estimated to "leak" from MSW at a rate of 1.13–2.24 MMT per year, based on 2016 estimates. This includes domestic leakage as well as mismanagement of exported waste (plastic scrap) by the United States to other countries. Comparing mismanaged plastic waste from other countries, Law et al. (2020) concluded that the United States was the 3rd to 12th largest contributor of plastic waste into the coastal environment with 0.51–1.45 MMT in 2016.

Not all waste, or plastic waste, leaks from the waste management system equally. Surveys and community science efforts (at large scales) have shown that plastics make up a large percentage of what ends up in the environment (70–80%), with the majority of plastic items being single-use, including packaging, as well as tobacco-related (e.g., cigarette filters, product packaging, and e-cigarette cartridges) and unidentified fragments sourced from larger items.

Conclusion 2: Materials and products could be designed with a demonstrated end-of-life strategy that strives to retain resource value.

Conclusion 3: Effective and accessible solid waste management and infrastructure are fundamental for preventing plastic materials from leaking to the environment and becoming ocean plastic waste. Solid waste collection and management are particularly important for coastal and riparian areas where fugitive plastics have shorter and more direct paths to the ocean.

Conclusion 4: The United States has a need and opportunity to expand and evolve its historically decentralized municipal solid waste management systems, to improve management while ensuring that the system serves communities and regions equitably, efficiently, and economically.

Conclusion 5: Although recycling will likely always be a component of the strategy to manage plastic waste, today's recycling processes and infrastructure are grossly insufficient to manage the diversity, complexity, and quantity of plastic waste in the United States.

Recommendation 1: The United States should substantially reduce solid waste generation (absolute and per person) to reduce plastic waste in the environment and the environmental, economic, aesthetic, and health costs of managing waste and litter.

PHYSICAL TRANSPORT AND PATHWAYS TO THE OCEAN

The ocean is Earth's ultimate sink, lying downstream of all activities. Almost any plastic waste on land has the potential to eventually reach the ocean. Major paths of plastics to the ocean are summarized in Figure S.3. These include urban, coastal, and inland stormwater; treated wastewater discharges; atmospheric deposition; direct deposits from boats and ships;

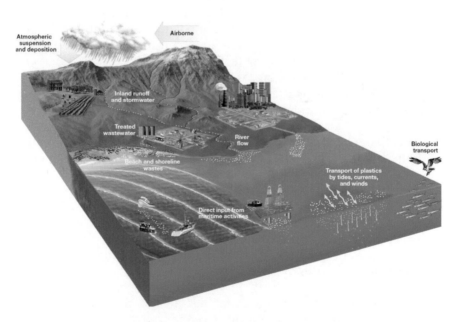

FIGURE S.3 Major transport pathways for plastics from land to the ocean.

beach and shoreline wastes; and transport from inland areas by rivers and streams.

Waterborne Pathways

The presumptive largest path of plastic mass from land to the ocean is from rivers and streams moving plastic wastes from inland and coastal areas to the sea. Rain and snowmelt flow over impervious surfaces such as paved streets and parking lots, carrying pollutants, including plastics, either into urban and stormwater systems that discharge to local areas or directly into rivers, streams, lakes, and coastal waters. Studies conducted in California indicated that the highest rates of plastic waste generation and loading were from industrial, retail, and residential areas, as well as highways and expressways.

Urban and suburban sewer flows to wastewater treatment plants are a smaller contributor of plastics to rivers or near-shore environments. They carry appreciable quantities of microplastics shed from clothing and other textiles. In wastewater treatment plants, most plastics are removed and concentrated in wastewater sludges that are buried in landfills or spread on land.

Other Pathways for Plastic Waste: Wind and Direct Input

As with water bodies, plastic items including everyday litter, such as bags and wrappers, large debris mobilized in severe windstorms, and microplastics can be suspended in the atmosphere and transported. Plastics also can be directly deposited into the ocean through losses of fishing and aquaculture gear, recreational gear (e.g., during boating or scuba diving), overboard litter, unregulated direct discharge, and cargo lost from ships and barges. Additionally, major storm events, such as hurricanes, floods, or tsunamis, can deposit massive amounts of debris in a relatively short period.

Challenge of Estimating Plastics Entering the Ocean

Although there is a fair understanding of the major mechanisms that transport plastic wastes to the ocean, it is difficult to make aggregate estimates of plastic fluxes to the ocean. A challenge in assessing paths and quantitative transport of plastics to the ocean is the limited number of quantitative studies and the variety of methods used and data reporting within the scientific community.

Conclusion 6: Regular, standardized, and systematic data collection is critical to understanding the extent and patterns of plastic waste inputs to the environment, including the ocean, and how they change over time.

DISTRIBUTION AND FATES OF PLASTIC WASTE IN THE OCEAN

The input of plastic waste in the ocean, as well as the Laurentian Great Lakes, is a reflection of the amount and type of plastic waste that enters the environment from a diversity of sources as well as the efficiency of the transport of this waste from upstream locations to the ocean and lakes. Its distribution and fate in the ocean are a reflection of transport by ocean currents and surface winds, and the degradation of plastics in the ocean. Plastic waste is found throughout the ocean including on coastlines and in estuaries, in the open ocean water column, on the seafloor, and in marine biota (Figure S.4).

Plastic Waste on Shorelines and Estuaries

Coastlines, including sandy beaches, rocky shorelines, and estuarine and wetland environments, are the recipients of plastic waste that may be generated locally, carried from inland sources, or brought ashore by storms, tides, or other nearshore processes. Items carried ashore may have been locally generated items that were trapped in the coastal zone or items generated elsewhere that were transported long distances. In 2018,

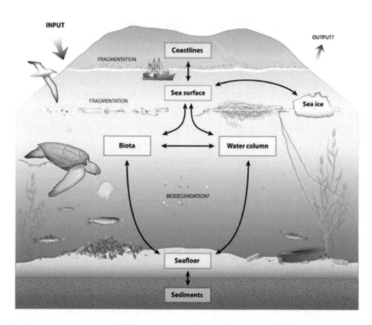

FIGURE S.4 Schematic of plastic waste in the ocean and interactions that can occur from land to sea and from surface to seafloor. SOURCE: Law (2017).

more than 32 million individual items were collected and categorized from more than 24,000 miles of beaches around the globe in the International Coastal Cleanup (Ocean Conservancy). The Top 10 list (highest number of items collected) has included the same consumer products year after year, including cigarette filters, food wrappers, beverage bottles and cans, bags, bottle caps, and straws.

Regional differences in amounts and trends of coastal debris are driven in part by debris source characteristics such as population size, land use, and degree of fishing activity. The state of Hawaii is particularly well known for suffering a disproportionately heavy marine debris burden, not only from locally based marine litter but also due to the state's mid-Pacific Ocean location and associated exposure to widely circulated plastic pollution originating throughout the Pacific Rim. Like Hawaii, Alaska coastlines are also a reservoir for significant amounts of plastic debris, which is often characterized by large, buoyant objects such as lines, buoys, and fishing nets.

Several major estuaries and inland freshwater waterways in the United States have been surveyed for plastic debris, especially microplastics in the water column or buried in sediments. These studies are widespread geographically—carried out in California, the Pacific Northwest, and along the eastern seaboard from New York to Florida, as well as in regions far from the ocean (Illinois, Montana, Wyoming, Wisconsin, western Virginia). Although a relatively small fraction of estuaries and rivers have been studied, the presence of microplastics in every study indicates that this waste is ubiquitous.

Plastic Waste in the Ocean Water Column and on the Seafloor

Sampling on the ocean's surface has allowed scientists to assess the large-scale accumulation of floating debris across ocean basins, which occurs in ocean gyres in both the northern and southern hemispheres. These accumulation zones, commonly referred to as "garbage patches," are mainly composed of microplastics that have broken apart from larger items, although large floating debris (especially derelict fishing gear, including nets, floats, and buoys) is also found.

Contrary to common misperceptions of "garbage patches," floating plastic debris is not aggregated together in a single large mass in the subtropical gyres and is instead dispersed across an area estimated to be millions of square kilometers in size. Even within the accumulation zones, particle concentrations (measured using plankton nets) can vary by orders of magnitude across spatial scales of tens of kilometers or less.

Microplastics, and occasional larger items such as plastic bags, have also been detected in the water column between the surface of the water and the seafloor. Vertical mixing of the water column driven by wind energy can distribute buoyant plastics to depths of tens of meters or greater, and interactions with organic matter and biota may also cause initially buoyant particles to become dense enough to sink. Macroplastics and microplastics have been found in seafloor (benthic) environments around the world. Observed concentrations vary greatly, both in the oceans and Laurentian Great Lakes, suggesting that proximity to sources, movement by water currents, and seafloor topography can act as concentrating mechanisms.

Impacts on and Distribution by Marine Life

Plastic waste has two especially well-studied impacts on marine and freshwater life: entanglement in plastic waste and ingestion-egestion of plastic waste. Ingestion is the taking in or consuming of food or other substances into the mouth or body. Egestion is discharging or voiding undigested food or other material, such as through feces or vomiting. One review by Kühn and van Franeker (2020) found documented cases of entanglement or ingestion by marine biota in 914 species from 747 studies—701 species having experienced ingestion and 354 species having experienced entanglement. Ingestion of plastic waste occurs at spatial scales ranging from the planktonic ingestion of micro- and nanoplastics to the ingestion of all sizes of plastic debris by whales (Kühn and van Franeker 2020, Santos, Machovsky-Capuska, and Andrades 2021). Microplastics in particular are ingested by marine biota and may move through the food web, ultimately to humans, but there is limited knowledge of effects throughout the food web and to humans specifically. Entanglement of marine life in ocean plastic waste is harmful or even deadly and may distribute this pollution via the active or passive movement of living or dead entangled organisms across aquatic habitats, though the frequency and ramifications of this mode of plastic waste distribution and transport are essentially unstudied. In addition to entanglement or ingestion, plastics are also colonized by microbes, and these microbial communities may serve as disease or pollutant vectors.

Transformation

Two main mechanisms are involved in the transformation and ultimate fate of plastics in the ocean: chemical and physical degradation. Physical degradation involves the breakage of bulk pieces of plastic. Chemical degradation involves the breakage of chemical bonds in the plastic structure and may be accelerated by exposure to ultraviolet radiation, high temperatures, and elevated humidity. Biodegradation of plastics by microbes has been proposed as a third mechanism, but measurable biodegradation (complete carbon utilization by microbes) in the environment has not been observed.

Conclusion 7: Without modifications to current practices in the United States and worldwide, plastics will continue to accumulate in the environment, particularly the ocean, with adverse consequences for ecosystems and society.

TRACKING AND MONITORING SYSTEMS

Documentation of the extent and character of plastic waste and potential sources or hotspots (reservoirs and sinks) informs prevention, management, removal, and cleanup strategies. This report illustrates the limited, or absence of, data from which to inform and implement effective plastic intervention actions. To inform source reduction strategies and policies, a national-scale tracking and monitoring program (or system of systems) is needed that spans the plastic life cycle (i.e., from plastic production to leakage into the ocean). No comprehensive life-cycle tracking and monitoring of ocean plastic waste presently exists. Tracking and monitoring systems currently in place focus on solid waste management inputs and plastic waste items detected in the environment and ocean. Tracking and monitoring play a critical role in evaluating the effectiveness of any interventions or mitigation actions, such as source reduction strategies or policies.

Role of the National Oceanic and Atmospheric Administration Marine Debris Monitoring and Assessment Project

The Marine Debris Monitoring and Assessment Project (MDMAP) is the flagship community science initiative of the National Oceanic and Atmospheric Administration Marine Debris Program that engages partner organizations and volunteers to foster a national shoreline monitoring program in support of research, science-based policies, and prevention efforts. The MDMAP surveys and records the abundance and types of marine debris on shorelines. To date, there are 9,055 surveys at 443 sites that span 21 U.S. states and territories and nine countries. Studies have demonstrated the utility of MDMAP data to estimate marine debris abundance and temporal trends, while also identifying associated limitations in spatial and temporal coverage, site selection, and variability among participants. A key shortcoming is the lack of a comprehensive national baseline for debris densities along the coast that hinders the ability to monitor change in general.

Vision for U.S. Marine Debris Tracking and Monitoring

A single, national U.S. marine debris (or plastic waste) tracking and monitoring system does not exist, nor does such a system appear to be feasible given the complexity of plastic production, use, and disposal (including leakage) and the diversity of environments through which plastics are transported and distributed. Furthermore, the specific aims of local, regional, national, and international efforts require the application of tracking and monitoring tools and technologies effective at particular spatial and temporal scales. However, the use of multiple, complementary tracking and monitoring systems in a synergistic approach implemented at sufficient spatial and temporal scales would contribute to (1) understanding the scale of the plastic waste problem and (2) identifying priorities for source reduction, management, and cleanup and assessing progress in reducing U.S. contribution to global ocean plastic waste.

The following describes tracking and monitoring systems of plastic waste items expected to have the greatest efficacy in ultimately reducing plastic waste inputs to aquatic systems. The specific type or types of plastic waste addressed by any system, including polymer types, associated chemicals, or other characteristics or parameters of interest, will necessarily reflect the aims and drivers of those entities establishing the tracking and monitoring system.

- Tracking and monitoring systems that are **scientifically robust, hypothesis-driven, and conceptualized** *a priori* **to answer critical knowledge gaps,** rather than approaches applied *post hoc* to plastic waste tracking and monitoring questions.
- **Technologically adaptive tracking and monitoring systems** that are able to incorporate and utilize current and emerging technologies to improve the spatial and temporal resolution of mismanaged plastic waste including the application of
  - remote sensing, autonomous underwater/remotely operated vehicles, sensor advances, passive samplers, and others;
  - crowdsourcing apps;
  - barcode tracking for recyclability and traceability;
  - biochemical markers and tracers that provide information on organismal exposure to environmental plastics, including legacy exposure and that which relates to organismal, including human, health; and
  - other current or emergent technologies.
- Tracking and monitoring systems that are **applied with sufficient spatial and temporal resolution** to capture meaningful data concerning knowledge and policy needs. For example, monitoring from

a watershed perspective or including **pre- and post-intervention tracking and monitoring to assess progress.**

- Tracking and monitoring systems that **collect data that are comparable and, when scientifically robust, compatible with prior efforts. Examples include using standardized** measurement units or experimental design.
- Tracking and monitoring systems that **leverage, rather than separate, U.S. federal investment** in the reduction of mismanaged plastic waste among government departments and create synergies in the federal response to such waste.
- Tracking and monitoring systems that **encompass the full life cycle of plastics,** thereby achieving an understanding of the "upstream" plastic waste compartments and associated leakages.

Recommendation 2: The National Oceanic and Atmospheric Administration (NOAA) Marine Debris Monitoring and Assessment Project, led by the NOAA Marine Debris Program, should conduct a scientifically designed national marine debris shoreline survey every 5 years using standardized protocols adapted for relevant substrates. The survey should be designed by an ad hoc committee of experts convened by NOAA in consultation with the Interagency Marine Debris Coordinating Committee, including the identification of strategic shoreline monitoring sites.

Recommendation 3: Federal agencies with mandates over coastal and inland waters should establish new or enhance existing plastic pollution monitoring programs for environments within their programs and coordinate across agencies, using standard protocols. Features of a coordinated monitoring system include the following:

- Enhanced interagency coordination at the federal level (e.g., the Interagency Marine Debris Coordinating Committee and beyond) to include broader engagement of agencies with mandates that allow them to address environmental plastic waste from a watershed perspective—from inland to coastal and marine environments.
- Increased investment in emerging technologies, including remote sensing, for environmental plastic waste to improve spatial and temporal coverage at local to national scales. This will aid in identifying and monitoring leakage points and accumulation regions, which will guide removal and prevention efforts and enable assessments of trends.

PRIORITIZED KNOWLEDGE GAPS

The committee identified the following knowledge gaps that impeded the ability to produce a complete assessment of the quantification of the U.S. contribution to global plastic waste requested in the statement of task.

Production: Limited access to transparent data on plastic production is a significant barrier to understanding the amounts and trends in quantities and types of plastic resins, a starting point for understanding how much may become waste.

Waste Management: There are not many national-scale data sets to understand sources, types, and relative scale of plastic waste generated and disposed or leaked to the environment beyond MSW data in the United States.

Transport and Pathways: A comprehensive understanding of the contribution of various transport pathways to plastics in the ocean is hindered by the complexity of the transport processes and the data needed to measure and model variability in fluxes over space and time. Improved understanding of the absolute and relative contributions of each pathway to plastics in the ocean could inform and prioritize actions to reduce the transport of plastics to the ocean.

Distribution and Fate: There is insufficient information to create a robust (gross) mass budget for marine plastics and their distribution in ocean reservoirs. To improve understanding of distribution and fate of plastics in the ocean, research is needed on the following issues:

1. The rate at which plastics degrade at various depths in the ocean, and how this varies by polymer type.
2. The fate of plastics in marine biota, including residence time, digestive degradation, and egestion and excretion rates.

Tracking and Monitoring: Currently, data collected by various monitoring efforts are not well integrated. There would be significant value in developing a data and information portal by which existing and emerging marine debris/aquatic plastic waste data sets could be integrated to provide a more complete picture of the efforts currently tracking plastic pollution across the nation. Such a portal would need to be supported by (1) standardized methods of data collection and (2) support for long-term data infrastructure. The ability to visualize the data contained in the portal would greatly enhance its utility for the public and decision makers to inform and assess the progress of plastic waste reduction efforts.

INTERVENTIONS TO REDUCE GLOBAL OCEAN PLASTIC WASTE

Despite limitations in complete quantification of plastic waste to the ocean, it is clearly ubiquitous and increasing in magnitude.

There is no one solution to reducing the flow of plastic waste to the ocean. However, a suite of actions (or "interventions") taken across all stages of the path from source to ocean could reduce ocean plastic waste and achieve parallel environmental and social benefits (Figure S.5). Taking systemic action across the plastic life cycle is necessary to avoid the current mismatch between how, and from what sources, plastic products are generated, and the waste and management systems that seek to control or limit the waste they produce. Choices of interventions within a systemic approach can help overcome limitations of each intervention. Actions to reduce ocean plastic wastes at each stage have different effectiveness and costs but together could constitute a regional, national, or global strategy for managing plastic wastes in the ocean and the environment. A policy challenge is to organize and implement a portfolio of interventions along this chain of plastic use and management to most effectively reduce or eliminate plastic wastes entering the ocean in light of both benefits and costs.

U.S. Federal Strategy for Reducing Plastic Waste

Although the United States lacks a nationwide systemic strategy for reducing plastic waste at all stages of the plastic waste cycle, many other countries (and some states) have been taking steps to address the plastic

Interventions

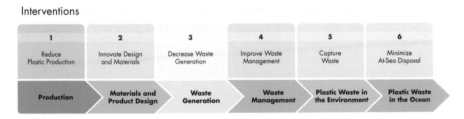

FIGURE S.5 Flow diagram of potential plastic waste interventions from plastic production to direct input into the ocean. SOURCE: Modified from Jambeck et al. (2018).

waste problem. As of 2018, 127 out of 192 countries regulated plastic bags restricting free retail distribution, and 63 countries mandated extended producer responsibility for single-use plastics, including deposit refunds, product take-back, and recycling targets. In addition, the European Union, Canada, and China have established systemic national goals and strategies designed around system-wide interventions.

The United States could similarly design and implement a coherent portfolio of effective and system-wide interventions by using a strategy and implementation plan that builds on existing efforts and adopts new models. Such a system could provide multiple benefits by (1) creating a clear policy or legal framework for reducing plastic waste in the ocean, (2) creating economic incentives toward reduction through reuse and recycling and away from production, (3) filling "leaks" in the U.S. waste management and pollution control systems, and (4) addressing funding gaps and reversing inequitable cost burdens.

Creating a framework for a system of interventions can align the United States with an emerging global approach. Moreover, a U.S. leadership role would help to position the nation to shape and influence global scale requirements around production, formulation, design, innovation, and waste reduction. This, in turn, can create innovation and economic opportunities that also internalize economic externalities and increase societal and environmental well-being.

> **Recommendation 4:** The United States should create a coherent, comprehensive, and crosscutting federal research and policy strategy that focuses on identifying, implementing, and assessing equitable and effective interventions across the entire plastic life cycle to reduce U.S. contribution of plastic waste to the environment, including the ocean. This strategy should be developed at a high level with a group of experts (or external advisory body) by December 31, 2022, and its implementation assessed by December 31, 2025. Such a strategy would enhance U.S. leadership in creating solutions to global plastic pollution and shaping modern industrial plastic policy.

1

Introduction

Global ocean plastic waste originates from materials introduced in the 20th century to deliver wide-ranging benefits (Thompson et al. 2009). Plastics increased an era of disposability for products and packaging used for a short time and then thrown away. The result has been a dramatic rise in plastic waste, some of which leaks to the environment, including the ocean. Plastic waste has a range of adverse impacts, some of which are only beginning to be recognized and understood (MacLeod et al. 2021). Over the past decade, research on ocean plastic pollution has revealed that plastic waste is present in essentially almost every marine habitat, from the ocean surface (van Sebille et al. 2020) to deep-sea sediments (Barrett et al. 2020) and the ocean's vast mid-water region (Choy et al. 2019). It also affects marine animals, including commercially important species of seafood, and ultimately humans (Barnes et al. 2009, Choy et al. 2019, Lusher et al. 2015, Santos, Machovsky-Capuska, and Andrades 2021).

The increasing visibility and scale of harmful effects of plastic pollution—from large items to microplastics—in freshwater and marine systems, along with related social and economic impacts, has brought the problem and the need for solutions to the forefront of public opinion and government concern. Global calls to action from all levels of government, the United Nations, civil society, and industry are translating to goals and plans of action at the national and international levels.[1] Local, state, and

---

[1]See https://www.gpmarinelitter.org/what-we-do/action-plans.

17

federal governments are simultaneously testing new policies and laws in response to public concerns. Society is grappling with the massive and increasing scale of global plastic waste: beach cleanups, local bans, extended producer responsibility schemes (Abbott and Sumaila 2019), "circular economy" commitments (Ellen MacArthur Foundation 2017), country-level plans and commitments (European Commission 2018, 2020), and calls for a global treaty (CIEL 2020, Karasik et al. 2020).[2]

The urgency has also prompted explosive growth in research, pilot approaches, and technology innovation globally. These efforts are moving forward quickly and will continue to provide new information and insights after the release of this report. Decision makers are calling for reliable syntheses of scientific knowledge and of global and national data (Environment and Climate Change Canada and Health Canada 2020). This report is intended to provide such an assessment. Definitions of key terms used in this report are found in Box 1.1.

STUDY CONTEXT

Since the invention of plastics in the 20th century, the production and use of plastics, and the volume of resulting plastic waste, have rapidly risen. The annual global production of plastics grew from about 2 million metric tons (MMT) in 1950 to 381 MMT in 2015 (Geyer, Jambeck, and Law 2017) and is projected to continue to increase (World Economic Forum, Ellen MacArthur Foundation, and McKinsey & Company 2016). Figure 1.1 depicts historic and projected plastic production growth, using numbers from Geyer, Jambeck, and Law (2017) and World Economic Forum, Ellen MacArthur Foundation, and McKinsey & Company (2016). Despite growing political and social will to mitigate plastic waste and reduce fossil fuel consumption, the plastic industry expects continued, unfettered growth of plastics demand and production over the next several decades (CIEL 2018). The figure does not include the COVID-19 pandemic and its effects on plastic consumption. However, historical trends reveal conditions revert to the pre-crisis trend (e.g., consumption levels after the 2007–2008 financial crisis). Box 1.2 provides a historical overview of the production and use of plastics.

Plastics are widely utilized throughout society because they have many diverse and useful properties for a broad array of applications. For example, plastics used in piping and other delivery system components help ensure water safety during transport, while plastic packaging extends food preservation and prevents contamination (Andrady and Neal 2009, Matthews, Moran, and Jaiswal 2021, Millet et al. 2018, Sharma and Ghoshal 2018). Compared to other packaging materials, such as glass,

---

[2]See https://usplasticspact.org/.

BOX 1.1
Key Terms Used in This Report

Plastics: A wide range of synthetic polymeric materials and associated additives made from petrochemical, natural gas, or biologically based feedstocks and with thermoplastic, thermoset, or elastomeric properties used in a wide variety of applications including packaging, building and construction, household and sports equipment, vehicles, electronics, and agriculture, and which occur in a solid state in the environment.

Virgin plastic: Plastic resin produced from a petrochemical, natural gas, or biobased feedstock, which has never been used or processed.

Solid waste: Residential, commercial, and institutional waste (Kaza et al. 2018). Industrial, medical, hazardous, electronic, and construction and demolition waste are excluded from this definition.

Plastic waste: Any plastic that has been intentionally or unintentionally taken out of use and that has entered a waste stream as part of a waste management process or released into the environment. Plastic waste in the environment is typically characterized according to size. Size classifications in this report follow the classifications used by the Joint Group of Experts on the Scientific Aspects of the Marine Environmental Protection (GESAMP) and adopted by the National Oceanic and Atmospheric Administration Marine Debris Program (GESAMP 2019).

Plastic solid waste: The subset of solid waste that is composed of plastics.

Marine debris or marine litter: Any persistent, manufactured, or processed solid material that is directly or indirectly, intentionally or unintentionally, discarded, disposed of, or abandoned into the marine, coastal, or Great Lakes environment. This definition excludes natural flotsam, such as trees washed out to sea, and focuses on non-biodegradable synthetic materials that persist in the marine environment (definition adapted from multiple sources).

Ocean plastic waste: A subset of marine debris; plastic waste in the marine environment including estuaries, coastlines, seawater (sea surface and water column), seafloor sediments, biota, and sea ice (these are similar ocean reservoirs as defined in Law 2017).
Ocean plastic waste, plastic marine debris, plastic marine litter, and marine plastic pollution are collapsed for clarity and used interchangeably.

Leakage: Loss of custodial control of plastic material to the environment, including during routine activities.

Microplastic: A plastic object from 1 to 1,000 μm in size as determined by the object's largest dimension (definition adapted from Hartmann et al. 2019).

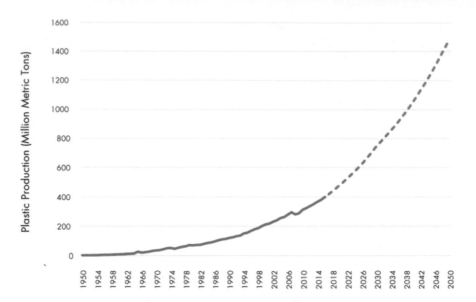

FIGURE 1.1 Global plastic production trend and projected growth. SOURCES: Data from 1950 to 2015 from Geyer, Jambeck, and Law (2017) supplemental material; projected numbers from Ellen MacArthur Foundation's annual industry growth estimations (World Economic Forum, Ellen MacArthur Foundation, and McKinsey & Company 2016)—2016–2020 has an annual 4.8% growth rate, 2021–2030 4.5%, and 2031–2050 3.5%. This does not include COVID-19 impacts.

plastic packaging uses less material, due to its strength, and less energy during transport, due to its lightweight nature (Andrady and Neal 2009, Millet et al. 2018). In construction, plastics are widely used because of their durability. Plastics used in medical settings have improved patient and worker safety (e.g., nitrile gloves, disposable syringes, and sterile products such as intravenous bags and dialysis tubes) and have been used to advance healthcare treatments (e.g., absorbable sutures, controlled drug delivery systems, orthopedics, hearing aids, artificial corneas, and prostheses) (Millet et al. 2018, North and Halden 2013).

The durability of plastics, and their resulting persistence in the environment, creates a particularly challenging ocean waste problem, as described below. At present, plastic waste is the least recycled and recyclable of all persistent solid waste (glass, metal) in the waste stream and the environment (Coe, Antonelis, and Moy 2019). Moreover, with population growth and consumption per capita increasing worldwide, plastics will continue to pollute the marine environment (Jambeck and Johnsen 2015, Jambeck et al. 2015).

BOX 1.2
A Brief History of the Production and Use of Plastics

The first fully synthetic plastic, Bakelite or Baekelite ($C_6H_6O \cdot CH_2O)_n$, was developed in 1907 by a researcher looking for a replacement for shellac. Leo Baekeland, the Belgian-American chemist responsible for developing Bakelite, is also credited with first employing the term "plastics" (Watson 2018). The material was patented in 1909 and marketed as a heat-resistant electrical insulator for radio and telephone housing, kitchen appliances, and other products.

Plastics are a class of solid synthetic or semi-synthetic materials based on long-chain organic polymers with high molecular mass and mostly linear structure. Polymers are formed when small molecules (monomers) combine chemically to form larger networks of repeating units. Some polymers occur naturally (e.g., rubber, proteins, DNA); others are synthesized.

Most synthetic plastic polymers today are derived from fossil hydrocarbons such as natural gas liquids or petroleum. Common polymers include polyethylene, polypropylene, polyester, and polystyrene. In 2019, 368 million metric tons of plastics were produced globally (Plastics Europe 2020). Plastic resins are produced primarily in North America, Europe, and Asia. Petrochemical plants convert fossil feedstocks into polymer resins. A key feature of plastics is that they can be molded, pressed, or extruded to form solid objects in a wide variety of shapes with a wide range of properties, including density, strength, and flexibility. Thermoplastic polymers form long, one-dimensional (linear) chains, and can be melted by heating and reformed. Thermosetting polymers undergo an irreversible chemical reaction when they solidify after the initial melting and cannot be melted and reformed. Plastics can be combined with additives, including colorants, fillers/reinforcements, flame retardants, plasticizers, and stabilizers, to change the properties of the material.

Plastics have become ubiquitous in packaging, building materials, clothing, automobiles and consumer products, medical devices, and many other applications. Plastics are versatile, inexpensive, easily mass-produced, durable, and light. Many of the characteristics that make them appealing in the modern global economy—low density, low cost, durability—become problematic when it comes to their disposal.

Understanding the Problem of Oceanic Plastic Waste

When plastics are taken out of use, whether intentionally or unintentionally, they become plastic waste. An estimated 8 MMT of plastic waste enters the world's ocean each year—the equivalent of dumping a garbage truck of plastic waste into the ocean every minute (Jambeck et al. 2015). Plastic waste that enters the ocean includes single-use items (designed to be used once before disposal, such as packaging, water bottles, or straws) and durable items. If current practices continue, the amount of plastics discharged into the ocean could reach up to 53 MMT per year by 2030, roughly half of the total weight of fish caught from the ocean annually (Borrelle et al. 2020, Jambeck and Johnsen 2015, Pauly and Zeller 2016).

The United States is a major contributor to global plastic waste: in 2016, the country generated an estimated 42 MMT of plastic waste—the largest mass of plastic waste generated by any country. The European Union (28 countries) generated the second highest amount of waste at 30 MMT, followed by India (26 MMT) and China (22 MMT) (Law et al. 2020) (Table 1.1).

Plastics deployed as "single-use" products or packaging, about 45% of the total produced each year, become plastic waste quickly, often within the year of manufacture (Geyer, Jambeck, and Law 2017). Other plastics remain in use for decades, sometimes repurposed from their original application. Eventually, all plastics are intentionally or accidentally "retired" from use and become waste.

Impacts of Oceanic Plastic Waste

Plastics have been lauded for their durability, convenience, and affordability. These same attributes make plastics a primary and pervasive environmental contaminant with widespread biological, ecological, and economic impacts (Andrady 2011, Beaumont et al. 2019, Mæland and Staupe-Delgado 2020, Wright, Thompson, and Galloway 2013). When plastics and plastic waste are inadequately managed, their impacts are seemingly as diverse as the types of plastic itself (Bucci, Tulio, and Rochman 2020). The full ramifications of our reliance on and exposure to plastics continue to be investigated.

Impacts of aquatic plastic waste range from entanglement and ingestion by marine life (Kühn and van Franeker 2020) to associated ecotoxicological effects on a wide variety of taxa (Anbumani and Kakkar 2018, Guzzetti et al. 2018), including humans (see Singh and Li 2012 as one example). Plastic waste also affects microbial ecology as microplastics in wastewater treatment plants have been shown to enrich antibiotic resistance genes and serve as a vector for human and wildlife pathogens (Pham, Clark, and Li 2021). Exposure to marine plastic waste via seafood is likely to be greater for populations that depend heavily on seafood for nutrition. The contributions of environmental plastic waste to blue carbon—carbon captured by the oceans, marine plants and algae, and coastal ecosystems—and impacts on blue carbon sinks relating to biogeochemical cycling and climate change warrant further attention. Finally, the nexus between plastics (production, use, and waste) and socioeconomic factors has varied direct and indirect effects. One example is the ecosystem devaluation and loss of tourism from increased marine debris (Leggett et al. 2014, 2018). Orange County, California would add $137 million to recreational expenditures and the regional economy if it reduced marine debris to zero. Conversely, if marine debris doubles, it would cost

TABLE 1.1 Plastic Waste Generation Values Across Countries

Country	Plastic Waste Generation (metric tons)	Total Waste Generation (metric tons)	% Plastic in Solid Waste	2016 Population (millions)	Per Capita Plastic Waste Generation (kg/year)
United States[a]	42,027,215	320,818,436	13.1	323.1	130.09
United States	34,020,748	263,726,732	12.9	323.1	105.30
EU-28[b]	29,890,143	243,737,466	11.7	511.2	54.56
India	26,327,933	277,136,133	9.5	1,324.5	19.88
China	21,599,465	220,402,706	9.8	1,378.7	15.67
Brazil	10,675,989	79,081,401	13.5	206.2	51.78
Indonesia	9,128,000	65,200,000	14.0	261.6	34.90
Russian Federation	8,467,156	59,585,899	14.2	144.3	58.66
Germany	6,683,412	51,410,863	13.0	82.3	81.16
United Kingdom	6,471,650	32,037,871	20.2	65.6	98.66
Mexico	5,902,490	54,151,287	10.9	123.3	47.86
Japan	4,881,161	44,374,189	11.0	127.0	38.44
Thailand	4,796,494	27,268,302	17.6	69.0	69.54
Korea, Rep.	4,514,186	18,576,898	24.3	51.2	88.09
Italy	3,365,130	29,009,742	11.6	60.6	55.51
Egypt, Arab Rep.	3,037,675	23,366,729	13.0	94.4	32.16
France	2,929,042	32,544,914	9.0	66.9	43.81
Pakistan	2,731,768	30,352,981	9.0	203.6	13.42
Argentina	2,656,771	18,184,606	14.6	43.6	60.95
Algeria	2,092,007	12,378,740	16.9	40.6	51.59
Malaysia	2,058,501	13,723,342	15.0	30.7	67.09
Spain	1,832,533	20,361,483	9.0	46.5	39.42

[a]Refined estimate for the United States.
[b]EU-28 countries are reported collectively.
SOURCE: Law et al. (2020).

Orange County $304 million (Abt Associates 2019). Importantly, many of these socioeconomic impacts disproportionately affect marginalized communities and are recognized as environmental justice issues (see Box 1.3 for more information, UNEP 2021b).

Environmental and Human Health Impacts

Exposure to the jarring, tragic images of iconic megafauna entangled in marine debris is, for many, their introduction to, and remains synonymous with, the ocean plastic waste problem. As early as the 1970s, entanglement in, and ingestion of, marine debris by ocean life was widely observed and recognized as an emerging concern (Laist 1997, Shomura and Yoshida 1985). Presently, 914 species are known to have entanglement or ingestion records (Kühn and van Franeker 2020).

Plastic waste interferes with animal health when it is mistaken for food or is incidentally consumed during feeding activities (see Santos, Machovsky-Capuska, and Andrades 2021 for a recent review). It can range from large plastic pieces ingested by whales to microplastics ingested by organisms of all sizes (Kühn and van Franeker 2020, López-Martínez et al. 2021). How plastic exposure, via ingestion or other routes, affects organisms is a subject of ongoing research. As one example, interactions of corals with plastics have shown reduced growth (Reichert et al. 2018), impaired feeding (Savinelli et al. 2020), decreased fitness (Savinelli et al. 2020), and reduced calcification (Chapron et al. 2018), among many other negative outcomes (Rocha et al. 2020). Ingestion of plastics entrains plastic pollution in the food web, with potential for bioaccumulation in predators that consume plastic-contaminated prey.

Marine plastic waste can also impact services provided by ocean ecosystems, from provisioning services to carbon sequestration. For example, it is impairing the cycling of nutrients and the biological carbon pump, which negatively impacts the ocean's carbon sink capacity (Galgani and Loiselle 2021, Kumar et al. 2021, Shen et al. 2020, Villarrubia-Gómez, Cornell, and Fabres 2018). There is a wide range of processes by which this occurs. A few examples include marine plastic waste affecting phytoplankton photosynthesis (Galgani and Loiselle 2021, Shen et al. 2020); a thicker barrier hindering air-sea gas exchange (Galgani and Loiselle 2021); microplastics increasing the sinking rates of zooplankton fecal pellets, thereby altering the vertical flow of carbon and nutrients (Cole et al. 2016, Villarrubia-Gómez, Cornell, and Fabres 2018); and plastic particles accumulating on the seafloor and affecting long-term carbon storage (Villarrubia-Gómez, Cornell, and Fabres 2018).

BOX 1.3
The Issue of Environmental Equity

U.S. EPA defines environmental justice as "the fair treatment and meaningful involvement of all people regardless of race, color, national origin, or income, with respect to the development, implementation, and enforcement of environmental laws, regulations, and policies" (U.S. EPA 2021a). Like other forms of environmental pollution (Castellón 2021, Fernández-Llamazares et al. 2020, Saha and Mohai 2005), the negative impacts of plastic production, use, and waste are disproportionately experienced by vulnerable populations (or those who are historically disenfranchised) in the United States (Castellón 2021, Mizutani 2018) and abroad (UNEP 2021b). In the United States, communities of color (Black, Brown, and Indigenous communities) have experienced environmental pollution at higher rates than white communities (Bullard 2014, Mizutani 2018).

From exploration of oil to extraction to the disposal of plastic waste, there are aspects along the entire life cycle of plastics that have disproportionately harmful effects on marginalized communities, from the local level to the international level (e.g., Bai and Givens 2021, UNEP 2021b). Oil drilling and well fields have negatively impacted Indigenous peoples, globally, who rely on natural resources for subsistence as well as their livelihoods (O'Rourke and Connolly 2003). Oil and natural gas extracted from the land are then sent to refineries to be chemically processed in petrochemical facilities that affect the life quality and potentially the health of residents in communities surrounding the facilities (UNEP 2021b). Communities surrounding chemical processing facilities are known as fenceline communities and are often exposed to toxic pollution (White 2018). Fenceline communities in the United States are disproportionately made up of minoritized groups, including Black Americans, Latinos, and low-income populations (White 2018).

Similar to plastic production, plastic waste is also an environmental justice issue. Bullard et al. (2008) built on the 1987 Toxic Wastes and Race report (United Church of Christ 1987), the groundbreaking study that first correlated waste facility sites to demographic characteristics. More than 20 years from the initial report, Bullard et al. (2008) showed that low-income and communities of color still experienced disproportionate exposure to hazardous waste facilities. The Tishman Environment and Design Center (2019) report noted that 58 of 73 (or 79%) municipal solid waste (MSW) incinerators are situated in environmental justice (EJ) communities. The report defines EJ communities as communities comprising 25% or more people of color and/or impoverished people. Plastics make up roughly 13% of MSW and, when burned, release toxic pollutants, such as dioxins, furans, mercury, and polychlorinated biphenyls (GAIA 2019, Tishman Environment and Design Center 2019, Verma et al. 2016). Even in small quantities, these pollutants have serious consequences on human health, including increased heart disease risk; intensified respiratory illnesses such as asthma and emphysema; increased rashes, nausea, or headaches; and impaired nervous system (Verma et al. 2016). Intensified respiratory illnesses have further implications for COVID-19, an illness that affects those with impaired respiratory systems more seriously.

Internationally, advanced economies externalize the cost of waste management by exporting plastic waste to less advanced economies (Bai and Givens 2021), who ultimately bear the brunt of the economic, social, and environmental costs of plastic waste (GAIA 2019). Before 2018, the United States exported most of its plastic waste to China. After China banned most plastic waste imports, the United States diverted its exported waste to other Southeastern Asian

BOX 1.3 Continued

countries—namely, Malaysia, Indonesia, and Thailand—although the amount exported decreased significantly (INTERPOL 2020). Because many communities have single-stream recycling in the United States, items are often disposed of improperly (UNEP 2021b). This improperly disposed of plastic waste (e.g., non-recyclable plastics and other harmful chemicals leached by certain plastics) is harmful to both environmental and human well-being in the low- and middle-income import countries (Bai and Givens 2021). The increased plastic waste imports in Southeastern Asia have resulted in increased burning of trash, illegal disposal, and unregulated recycling operations (GAIA 2019). This has had broad impacts, including polluted water stores, crop loss, respiratory ailments from burning activities, and organized crime in regions most impacted by the increased plastic waste imports (GAIA 2019).

The impacts from U.S.-generated plastic waste on its residents, humankind, and the environment, including the global ocean, are substantial. In this era of intense globalization, the direct and indirect causes of environmental harm are often entangled in complex structures involving local groups, state authorities, international bodies, and corporate institutions (Davies 2018).

Some effects of plastic ingestion may be attributed to chemicals used to manufacture plastics, which can leach from plastics into animal tissues (Engler 2012, Jarosova et al. 2009, Koelmans, Besseling, and Foekema 2014, Teuten et al. 2007). Leaching of chemicals may vary by plastic type, weathering of plastics in seawater, or by reactions with digestive fluids. By 2010, more than 120 scientific studies on the role of plastics and their additives in human and animal health—largely through these compounds' actions as endocrine disrupters—had been published (Halden 2010). From animal studies, endocrine-disrupting effects from plastics-associated compounds, including reproductive disease, sperm epimutations, and obesity, have been found to transmit to offspring (Manikkam et al. 2013). Recently, microplastics have been found in human placentas examined after birth, despite a plastic-free birthing protocol (Ragusa et al. 2021).

Adsorption of exogenous chemicals, metals, and persistent organic pollutants on plastic litter also introduces toxins to the food web when plastics are ingested, although mechanisms and quantities of transfer and their impacts are still being investigated (Amaral-Zettler, Zettler, and Mincer 2020, Kögel et al. 2020, Mato et al. 2001, Rios, Moore, and Jones 2007, Rochman et al. 2013, Rochman, Hentschel, and Teh 2014, Saliu et al. 2019, 2021, Santana-Viera et al. 2021, Teuten et al. 2007, Wright, Thompson, and Galloway 2013). Trophic transfer of microplastics through both juvenile and adult salmon predation on zooplankton containing plastics, for example krill and copepods, is estimated at up to 91 plastic particles daily (Desforges, Galbraith, and Ross 2015).

Economic Impacts

The true economic impact of global ocean plastic waste remains largely unknown, but work to date suggests the costs are substantial. The physical removal of coastal marine debris is costly (Stickel, Jahn, and Kier 2012), but these estimations do not routinely include nonmarket ecosystem service valuations or the depreciation of environmental services and resources. Economic impacts of plastic waste also do not include the costs associated with properly managing waste through the use and ultimate discard of the plastics manufactured.

Inextricably linked to ocean plastic pollution's impacts on individuals, communities, and species are its effects on ecosystems and its economic ramifications. One estimate places the economic damage to marine ecosystems from plastics at a minimum of $13 billion annually (UNEP 2018). Beaumont et al. (2019) show that plastics negatively affect the ability of the marine ecosystem to function fully and therefore reduce its ability to continue to provide marine ecosystem services such as provision of fisheries, carbon sequestration, cultural heritage, and recreation. The authors estimated that the economic cost of marine plastic pollution is $3,300 to $33,000 per metric ton of plastic waste already in the ocean per year.

Economic impacts of mismanaged plastic waste can also be estimated from studies of the ecosystem service values the plastic waste may impact. For example, the perceived value of a beach is intimately linked with its overall cleanliness (Leggett et al. 2018), and local plastic hotspots from river influx threaten water quality (Keswani et al. 2016). A study in California determined that removing 50–100% of the litter on Orange County beaches could yield California residents $67–$148 million during the 3 months of summer (Leggett et al. 2014). When nonmarket values are unaccounted for and the degradation of ecosystem services is not considered, there is a failure to comprehensively interpret the total economic value.

ORIGIN OF THIS STUDY

Research on marine plastic pollution has grown at an exponential rate in the past few years along with increased public, governmental, and legislative interest into the causes of plastic pollution and potential interventions. One legislative instrument was the Save Our Seas 2.0 Act, which was sponsored by a bipartisan group of 19 senators and passed into law on December 18, 2020, in the 116th Congress (Public Law Number 116-224). This law stipulates requirements and incentives to address marine debris and expands the reach of the first Save Our Seas Act (Public Law Number 115-265).

This study, among other studies called for in the Save Our Seas 2.0 law, examines U.S. contributions to global oceanic plastic waste. The study was sponsored by the National Oceanic and Atmospheric Administration's Marine Debris Program.

STUDY SCOPE AND APPROACH

This report focuses on those aspects of the uses of plastics and the oceanic waste they generate that are laid out in the statement of task for this study (Box 1.4). The rapid growth and evolution of the salient literature and the sheer scope of the issues involved required that the committee focus the report on the most pressing issues in need of attention.

Conversely, the statement of task does not cover all important topics on plastics, such as other Earth system components impacted by plastics, human and environmental impacts of ocean plastic pollution (including microplastics), sources and impacts of derelict fishing gear, detailed impacts of environmental equity, or impacts of land-based waste disposal or incineration methods. The scholarship on these areas is expansive and, where relevant, summaries and references to articles and reports on these topics are included in the text.

Chapter 2 discusses plastic production and global trade in the United States (statement of task [SOT] 1, 2a, 3a). Chapter 3 examines how plastic waste is managed (SOT 2a, 3b, 3c, 3d). Chapter 4 details the transport mechanisms of plastics and the pathways they encounter from source to the ocean (SOT 1, 2a, 2b). Chapter 5 starts off with an overview of global ocean plastic waste and then examines distribution and fate of plastics in the ocean, from estuaries to the open ocean (SOT 1, 2c). Chapter 6 considers tracking and monitoring systems (SOT 4a, 4b). Throughout Chapters 2–6, recommendations of prioritized knowledge gaps and means to reduce plastic waste are explored (SOT 5). Chapter 7 closes the report and provides intervention categories for how the United States might reduce global ocean plastic waste contributions (SOT 6).

BOX 1.4
Statement of Task for This Study

An ad hoc committee will be convened to undertake a study on the United States contributions to global ocean plastic waste.

1. Evaluate United States contributions to global ocean plastic waste, including types, sources and geographic variations.
   a. compare to global estimates of plastic waste entering the ocean
   b. assess US contribution by mass and percentage of total
   c. evaluate US contribution according to size class
2. Assess the prevalence of marine debris and mismanaged plastic waste in saltwater and freshwater United States waterways.
   a. Include contributions from land-based industry, littering, mismanaged waste, wastewater treatment plant discharge, river discharge, accidental transportation-related releases, or other significant sources
   b. evaluate how much and what proportion of upstream waste flows downstream to the ocean
   c. include state of knowledge about distribution and fate of different types of plastic within the water column, nearshore and offshore.
3. Examine the import and export of plastic waste to and from the United States, including the destinations of the exported plastic and the waste management infrastructure and environmental conditions of these locations.
   a. estimate U.S. virgin plastic shipped internationally for manufacture of plastic products in other countries
   b. determine the mass and percentage of United States total plastic waste exported (historic and current estimates) and how these estimates compare to other nations
   c. identify the origin of plastic materials in the US waste stream (plastic feedstock and manufactured products)
   d. assess the trend of landfill deposits and debris in US waterways following current plastic export bans to other countries
4. Assess the potential value of a national marine debris tracking and monitoring system and how such a system might be designed and implemented.
   a. consider how the tracking and monitoring system could be used to identify priorities for source reduction and cleanup, assess progress in reducing US contribution to global ocean plastic waste, and determine which existing systems or technologies would be most effective for reducing inputs of plastic waste to the ocean.
   b. assess how the Marine Debris Monitoring and Assessment Project protocols can inform a nationwide shoreline monitoring effort when implemented at greater spatial and temporal resolution
5. Develop recommendations on knowledge gaps that warrant further scientific inquiry.
6. Recommend potential means to reduce United States contributions to global ocean plastic waste.

2

Plastic Production and Global Trade

Plastic production operates at a global scale. As described in sub-sequent chapters, the United States contributes to the problem of global ocean plastic waste as a result of plastics produced and used in this country or exported to other nations, as well as from imported plastics manufactured elsewhere that enter the U.S. waste stream. This chapter describes the production of materials that may become plastic waste: feedstocks for plastic resins, the production process, biodegradability of plastics, the types of products generated from plastics, and characteristics that create challenges for the waste stream and our environment. The vast majority of plastics are produced from natural gas or petroleum feedstocks, with a small portion from biobased (renewable) feedstocks, resulting in implications for plastic production trends, potential impacts of production, and waste management.

PROPERTIES OF PLASTICS

Chemical Structure of Plastics

Of the world's thermoplastics (plastic polymers forming long, one-dimensional [linear] chains that can be melted by heating and reformed), 76.7% are hydrocarbon plastics (Law and Narayan 2022). Hydrocarbon plastics are polymers made from monomers composed of carbon and hydrogen (ISO 472:2013). They are carbon-carbon backbone polymers as shown in Figure 2.1. Examples of hydrocarbon plastics include linear

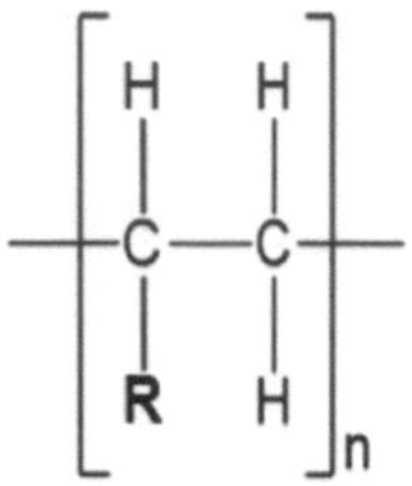

$$\left[\begin{array}{cc} \overset{\displaystyle H}{\underset{\displaystyle \mid}{}} & \overset{\displaystyle H}{\underset{\displaystyle \mid}{}} \\ -C & -C- \\ \underset{\displaystyle R}{\mid} & \underset{\displaystyle H}{\mid} \end{array} \right]_n$$

R = -H; Polyethylenes (HDPE, LDPE, LLDPE)

R = -CH₃; Polypropylenes (PPs)

R= -CH₂-C₆H₅; Polystyrene (PS)

R = -Cl; Poly(vinyl chloride) PVC

Carbon-Carbon backbone

FIGURE 2.1 This figure shows the general chemical structure of carbon-carbon backbone polymers. The C-C bond is in red and the H indicates the hydrogen atoms bonded to the C atoms. The R indicates a side group that varies among the specific plastic materials listed.

---

low-density polyethylene (LLDPE), low-density polyethylene (LDPE), high-density polyethylene (HDPE), polypropylene (PP), polystyrene (PS), and polyvinyl chloride (PVC) (Agamuthu et al. 2019, Law and Narayan 2022). The strong carbon-carbon bond makes these plastics resistant to biodegradation at a rate incompatible with timely removal from the environment. This resistance to biodegradation, together with plastics' lightweight and ubiquitous nature, results in persistence and accumulation of hydrocarbon plastics in natural environments.

Plastic Feedstocks

Synthetic plastics can be produced from fossil feedstocks or renewable biomass. Globally, more than 99% of plastics are produced from fossil feedstocks—petroleum (crude oil) or natural gas (British Plastics Federation 2019, CIEL 2017, 2020, Skoczinski et al. 2021).

Biobased plastics are plastics in which the carbon originates, in whole or in part, from renewable biomass feedstock such as sugar cane, canola, and corn. Biobased plastics are less than 1% of all plastics produced globally (European Bioplastics 2020). Biobased carbon content of a product is measured as the amount (mass) of biobased carbon as a percentage of total organic carbon (ASTM D6866, ISO 16620 series, USDA BioPreferred program).

Biodegradability of Plastics

Degradation, and specifically biodegradation, depends on the chemical and physical structure of the plastics and the characteristics of the receiving environment (e.g., industrial composting, soil, ocean, backyard composting), not from where the carbon originates. For example, hydrocarbon plastics (i.e., plastics with a carbon-carbon backbone) can be manufactured from biomass carbon feedstocks. These plastics are biobased, but they will have identical chemical structure as those manufactured using fossil carbon feedstocks and exhibit the same non-biodegradable, persistent characteristics.

The complex relationship between biobased plastics and biodegradability contributes to consumer and labeling confusion (IEA Bioenergy 2018, U.S. EPA 2020). Biobased refers to the plastic feedstock and does not relate to how biodegradable the plastic is (Closed Loop Partners 2020, Law and Narayan 2022). Several, but not all, biobased plastics are biodegradable and industrially compostable at end of life.

PLASTIC PRODUCTION

Over a 50-year period, global plastic production increased nearly 20-fold, from 20 million metric tons (MMT) in 1966 to 381 MMT in 2015 (Geyer, Jambeck, and Law 2017). Table 2.1 summarizes recent estimates of annual and cumulative production of plastics in the United States and globally. Approximately one-fifth (19%) of 2019 global plastic production occurred in North America, second to Asia (Plastics Europe 2020). Plastic production is projected to increase by 200% and 350% by 2035 and 2050, respectively (Geyer, Jambeck, and Law 2017, Lebreton and Andrady 2019, World Economic Forum, Ellen MacArthur Foundation, and McKinsey & Company 2016). More than 90% of plastics are made from virgin fossil feedstocks, which utilize roughly 6% of global oil consumption (World Economic Forum, Ellen MacArthur Foundation, and McKinsey & Company 2016).

With an estimated 3.5–3.8% annual growth rate (World Economic Forum, Ellen MacArthur Foundation, and McKinsey & Company 2016), plastics are projected to make up approximately one-third of oil demand growth in 2030 and almost half by 2050 (IEA 2018). The World Economic Forum and International Energy Agency recognize petrochemical, and particularly plastic, growth as an essential component in oil demand growth through 2050 (CIEL 2020). The fracking boom produced a surplus of cheap natural gas (CIEL 2020), and oil companies are strengthening and integrating petrochemical production and markets into their business models (IEA 2018). Oil and gas companies have invested more than $200 billion in plastic production (CIEL 2020) and intend to invest another $400 billion in virgin plastic production in the next 5 years (Bond et al. 2020, Brock 2020). By contrast, oil and gas companies will dedicate $2 billion to reducing plastic waste in the same period (Brock 2020).

TABLE 2.1 Recent Estimates of Annual and Cumulative Production of Plastics in the United States and Globally

Data Source	Annual Production		Cumulative Production Since ~1950	
	USA	Global	USA	Global
American Chemistry Council 2021b Includes HDPE, PVC, LDPE, LLDPE, PP, PS, EPS, TPU Excludes PET, thermosets, resin fibers	41 MMT in 2020[a]	–	[1,500–2,000] MMT	–
Plastics Europe 2020 Includes thermoplastics, thermosets, polyurethanes, elastomers, adhesives, coatings, sealants, PP fibers Excludes PET, PA, and polyacryl fibers	70 MMT in 2019[b]	368 MMT in 2019	–	–
Geyer, Jambeck, and Law 2017 Includes thermoplastics, thermosets, polyurethanes, elastomers, coatings, and sealants; polyester, polyamide, and acrylic fibers; additives	–	407 MMT in 2015	–	8,300 MMT in 2015

NOTE: Information about what is included and excluded from each estimate is taken directly from each data source. Square brackets indicate "on the order of" or "approximately." EPS = expanded polystyrene, HDPE = high-density polyethylene, LDPE = low-density polyethylene, LLDPE = linear low-density polyethylene, NAFTA = North American Free Trade Agreement, PA = polyamide, PET = polyethylene terephthalate, PP = polypropylene, PS = polystyrene, PVC = polyvinyl chloride, TPU = thermoplastic polyurethane.
[a]See Table 2.2 for definition of "domestic production," which may also include Canada and Mexico.
[b]Annual USA production is for NAFTA countries.

Importantly, the fossil fuel industry has benefited from continued tax subsidies for the past century. Historically, tax subsidies were necessary to incentivize new energy sources in the United States (Coleman and Dietz 2019). In 2015, the International Monetary Fund calculated U.S. energy subsidies to amount to $649 billion (IMF 2019), with 80% going to natural gas and crude oil (Coleman and Dietz 2019). Because plastics are made from fossil fuels, these tax subsidies greatly reduce the cost of fossil fuel feedstocks, making it a more profitable option for plastic production (CIEL 2018).

North American Plastic Production Trends

Data on U.S. plastic production was provided by the American Chemistry Council (ACC) (American Chemistry Council 2021b). ACC notes that the "data for all years may not truly be comparable, affecting the validity of growth rate calculations." In addition, there are variations within the data on what is considered "domestic" for each resin. No data for any resin used in this report are based only on U.S. production data. Instead, they include Canada or represent all of North America (United States, Canada, Mexico). ACC's methodology indicates that its reports cover 95–100% of total production in the United States/Canada. Furthermore, data are not available to show plastic usages by sector for the United States or North America. The data from ACC do not distinguish between fossil-based or biobased plastics.

Although individual thermoplastic resin production trends vary, data provided by ACC show that the overall trend for resin supply and production in North America has been increasing over the past 20 years (Figure 2.2). ACC does not report data on polyethylene terephthalate (PET), thermoset, and resin fibers, and the committee was unable to identify data for these materials. In 2019, Plastics Europe estimated 70 MMT

TABLE 2.2 Definition of "Domestic" for a Variety of Thermoplastic Production Data Provided by the American Chemistry Council (ACC)

Plastic Polymer (Abbreviation)	Definition of "Domestic"
Polyethylene terephthalate (PET)	ACC does not collect data on PET
High-density polyethylene (HDPE)	United States and Canada
Polyvinyl chloride (PVC or Vinyl)	United States and Canada
Low-density polyethylene (LDPE)	United States and Canada
Linear low-density polyethylene (LLDPE)	United States and Canada
Polypropylene (PP)	For 2001–2006: United States and Canada
	For 2007–2020: North American Free Trade Agreement (NAFTA)
Polystyrene (PS)	For 2001–2010: United States and Canada
	For 2011–2020: NAFTA
Expanded polystyrene (EPS)	United States and Canada
Thermoplastic polyurethane (TPU)	United States and Canada

SOURCE: American Chemistry Council (2021b).

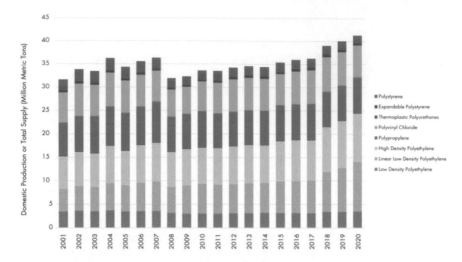

FIGURE 2.2 North American supply and production of plastic resin types from 2001 to 2020 by weight. Polystyrene and expandable polystyrene values are based upon total domestic supply, which includes productions and imports. All other resin values are based upon domestic production. Domestic indicates United States and Canada or United States, Canada, and Mexico, depending on the resin type and year (see Table 2.1). SOURCE: American Chemistry Council (2021a).

of plastic resins for North America, which accounts for 19% of global production (Plastics Europe 2020). Using ACC's data for eight types of resin, the committee estimates that a total of 41.1 MMT of plastic resins was produced in North America (Table 2.3). The PET Resin Association reports 2.8 MMT of PET production in North America and no specific year is noted (Heller, Mazor, and Keoleian 2020, PET Resin Association 2015).

For the eight groups of resin illustrated in Table 2.2, 41.1 MMT of plastics were produced in 2020 in North America. In 2020, both LLDPE and HDPE had production values at 10.4 MMT (Table 2.2). Plastic production trends in North America over the past two decades have varied for different plastic resins (Figure 2.3). LLDPE has been steadily increasing in domestic production over the past 20 years to reach 10.4 MMT in 2020 (Figure 2.3). For PVC, other than a dip in domestic production in 2008 and the following few years (likely due to a period of recession), production has remained consistent over the past 10 years. Around the same time period as the dip in production (2008), domestic sales decreased, but exports increased, keeping production relatively consistent (Figure 2.3). Total supply of PS has had an overall decreasing trend since

TABLE 2.3 Common Thermoplastic Resin Types, Associated Resin Codes, Their Predominant Uses, and Quantity of Resin Supply/ Production by Weight in Fiscal Year 2020 in North America as Reported by the American Chemistry Council by Plastic Type

Plastic Polymer (Abbreviation)	Resin Code	Uses	Million Metric Tons (MMT)
Polyethylene terephthalate (PET)	1	Single-use beverage bottles, food containers, textiles, etc.	Data not available
High-density polyethylene (HDPE)	2	Milk bottles, detergent bottles	10.4
Polyvinyl chloride (PVC)	3	Window frames, profiles, floor and wall coverings, pipes, cable insulation, garden hoses, inflatable pools, etc.	6.9
Low-density polyethylene (LDPE)	4	Single-use plastic bags, reusable bags, trays and containers, agricultural film, food packaging film, etc.	3.5
Linear low-density polyethylene (LLDPE)	4	Single-use plastic bags, reusable bags, trays and containers, agricultural film, food packaging film, etc.	10.4
Polypropylene (PP)	5	Food packaging, candy and snack wrappers, microwave containers/ dishware, pipes, automotive parts, non-woven textiles, personal protective equipment/masks, fishing gear and nets, etc.	7.8
Polystyrene (PS)	6	Food packaging (e.g., cups, utensils), electrical and electronic equipment, etc.	1.6
Expanded polystyrene (EPS)	6	Food packaging (to-go containers, coolers), building insulation, electrical and electronic equipment, inner liner for fridges, etc.	0.4
Thermoplastic polyurethane (TPU)	7 "other category"	Clothing (Spandex), home building, automotive, industrial products	0.1
Total (2020)			**41.1**

NOTE: All of these plastics have carbon-carbon backbones except for PET and TPU.

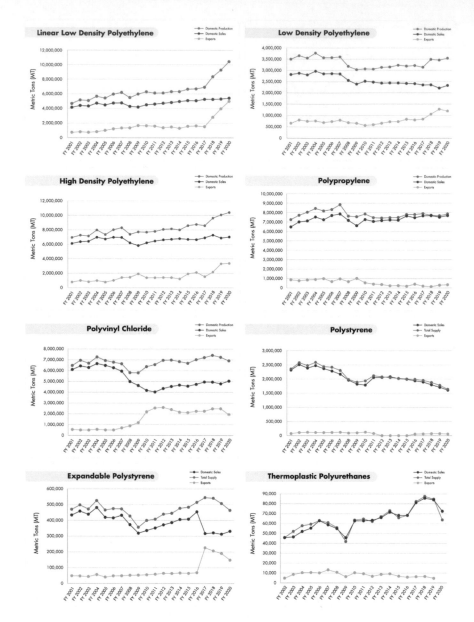

FIGURE 2.3 Domestic production (blue), sales (purple), and exports (green) from 2001 to 2020 for linear low-density polyethylene, low-density polyethylene, high-density polyethylene, polypropylene, and polyvinyl chloride. Total supply (blue), domestic sales (purple), and exports (green) from 2001 to 2020 for polystyrene and expanded polystyrene. Total supply (blue), domestic sales (purple), and exports (green) from 2001 to 2020 are illustrated for thermoplastic polyurethanes. Supply includes imports and domestic production. Domestic indicates United States and Canada or United States, Canada, and Mexico, depending on the resin type and year. More information is available in Table 2.2. SOURCE: Data from American Chemistry Council (2021b).

2005, with a consistent decrease over the past 10 years. Most of the supply is used domestically, with exports of PS being consistently relatively low (Figure 2.3).

COVID-19 Impacts on Plastic Use

While the COVID-19 pandemic has highlighted the importance of and the heavy reliance on single-use plastics in the medical field, the pandemic has significantly increased plastic usage—single-use plastics in particular—and associated waste across many aspects of daily life. This is in large part due to the unprecedented demand for personal protective equipment (PPE) for both healthcare workers and the average citizen, safety screens, and single-use plastics, such as shipping plastics, plastic bags, and restaurant takeout containers (De Blasio and Fallon 2022).

PPE is largely made from plastics. Surgical and N95 masks are commonly made of PP or PS (Henneberry 2020, Patrício Silva et al. 2021), although polycarbonate, polyethylene, polyester, polyurethane, and polyacrylonitrile are also used (Chellamani, Veerasubramanian, and Vignesh Balaji 2013). Polycarbonate is normally used for the production of visors (Roberge 2016), goggles, and glasses (Edwards 2020). Disposable medical gowns are made up of different synthetic fibers—including PP, polyester, and polyethylene—while reusable gowns are made from 100% cotton, 100% polyester, or a polyester/cotton blend (Kilinc 2015). As such, plastics have become an essential tool to protect against transmission of the COVID-19 virus (De Blasio and Fallon 2022, Dharmaraj et al. 2021).

PLASTIC TRADE

To assess U.S. exports and imports of plastics and plastic-containing goods, the U.S. Census Bureau Comtrade database was queried for the category "39. Plastics and articles thereof" (U.S. Census Bureau 2021). This classification system is based on coding for all commodities trading around the world. "Plastics and articles thereof" includes plastic polymers in primary forms, tubes, self-adhesive plates, baths, sinks, packaging goods, tableware, builders' plastics, and scrap waste, among other options (Comtrade 2020). This category does not include apparel and other clothing accessories (Codes 59–63) or toys and games (Code 95), which may also have plastic components.

Two data measures were downloaded—the customs value (gen), which is the average price paid per unit; and the card count, which is the number of individual import line items. This information was used to create Figures 2.4 and 2.5.

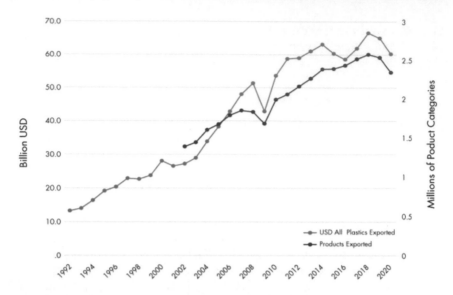

FIGURE 2.4 U.S. Census trade data indicating the value of plastic products exported and number of plastic product categories exported from 1992 to 2020. SOURCE: Data from U.S. Census Bureau (2021).

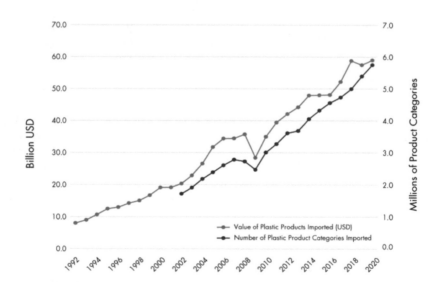

FIGURE 2.5 U.S. Census trade data indicating the value of plastics products imported and number of plastic product categories imported from 1992 to 2020. SOURCE: Data from U.S. Census Bureau (2021).

U.S. Exports

According to the U.S. Census Trade data, in 2020, the United States exported 2,342,368 categories of plastic products ("the number of individual export line items") at a value of $60.2 billion. This was a decrease from 2019 when 2,534,738 categories of plastic products were exported at a value of $65 billion. The trend of the USD value of plastic product exports and number of categories over the past nearly three decades have been increasing overall (Figure 2.4).

Although some domestic production of plastics is intended for export, it is unclear how increased plastic regulation around the globe will impact exports/imports (CIEL 2020). A UNEP (2018) report noted that 127 out of 192 countries studied have implemented some form of regulations to reduce plastic waste, including limiting plastic use. In particular, 61 countries, predominantly in Africa, have adopted manufacturing and import bans. As such, it is unclear if foreign markets can completely absorb the surplus supply of U.S. plastic production.

U.S. Imports

According to the U.S. Census Trade data, in 2020, the United States imported 5,747,472 categories of plastic products ("the number of individual import line items") at a value of $58.9 billion. This was an increase from 2019 when 5,390,001 categories of plastic products were imported at a value of $57.4 billion. In both years, the imports were a larger number of categories (more than double the categories that were exported), but the total value of imports was less than exports. The trend of the USD value of plastic product imports and number of categories over the past nearly three decades has been overall increasing (Figure 2.5).

These export and import data suggest that the United States plastic imports comprise more lesser value items and plastic exports comprise more higher value items.

Overview of Plastic Economics

The market price of plastics reflects their production cost, referred to in economics as a "direct cost." But it does not reflect indirect costs or benefits, such as the environmental and ecological costs (e.g., leakage of plastic waste into the environment, cost of proper disposal) or benefits (e.g., ecosystem services). Indirect effects, which impose costs or benefits on society and/or the environment that are not reflected in market prices, are referred to as "externalities" in economics. Failure to include those externalities in the price consumers pay for plastics may lead to greater reliance on plastics than is socially optimal. In order to account for externalities, market prices and benefit-cost analyses should expand and include more sources of valuation.

Kemp-Benedict and Kartha (2019) define value systems, value, and valuation:

> *Value systems* are normative and moral frameworks that guide action. Within their value systems, people assign *value* to actions or objects in the degree to which they meet user-specified goals, objectives, or conditions. *Valuation* is then the process by which values are assigned to actions and objects.

Historically, economic theory distinguishes between "use value" and "exchange value." Use value is the value of a commodity in use, such that it satisfies some human or societal need or desire. Exchange value is the value of a commodity in exchange, such that a commodity can be exchanged for something else. While classical economists would agree that ecosystem services have a high use value, their theory suggests that a commodity's exchange value is derived from the cost of labor to produce it. Because natural capital and ecosystem services are taken from nature at no cost, their exchange value is assumed to be zero (Kemp-Benedict and Kartha 2019). When natural capital is not clearly incorporated into economic decision making, ecosystem services are uncounted benefits (positive externalities), and the harms done to an ecosystem and its services are uncounted costs (negative externalities) (Kemp-Benedict and Kartha 2019). Environmental economic theory suggests a more complete accounting of values into a total economic value (Goulder and Kennedy 1997). Total economic value is the sum of market and nonmarket values, direct use value and indirect use value, option value, existence value, and bequest value.[1]

When all costs and benefits of using plastic products are taken into account, the net benefits of using plastics (i.e., the private value to people of using plastics, minus private production and waste disposal costs) are considered together with the environmental and human health externalities

---

[1]While exchange value and use value are the most commonly used distinctions of value, environmental economists recognize further distinctions among value, including market and nonmarket values; direct use value and indirect use value; option value; existence value; and bequest value. Market value is defined as the "balance between production costs and what people are willing to pay [while] non-market value is something that is not bought or sold directly" (GreenFacts 2021). Direct use value is consumptive and indirect use value is the "value of leaving something alone" (Kemp-Benedict and Kartha 2019). Option value is the value of postponing use to an indeterminate future. Existence value is the benefit people receive from knowing that a particular natural resource exists (e.g., Antarctica). Bequest value is the value of satisfaction people experience when natural resources are preserved or conserved.

of plastic production and plastic waste (Baumol and Oates 1988). The environmental costs of plastic waste are related to the amount and impacts of plastics that "leak" into the environment, including the ocean, related to total plastic production and plastic waste management (see Chapter 3). If economic assessments focus only on market value or private cost and do not consider externalities (positive or negative), it results in an incomplete understanding of marine plastic waste's economic impacts (Jambeck et al. 2020) and incomplete, distorted price information for consumers.

In terms of private production costs and benefits, plastics remain one of the world's most efficient and cost-effective classes of materials. Their properties can be modified to meet specific needs, and they can be molded into a variety of shapes and products (Hopewell, Dvorak, and Kosior 2009, OECD 2018, UNEP 2014). Because 99% of plastics are made from fossil-based feedstocks and the fossil fuel industry is subsidized, plastics are an artificially cheap commodity (CIEL 2018). This means that substitution with alternative, often more expensive materials such as concrete, wood, metal, and glass usually comes at a private cost (cost paid by the consumer or producer), and can lead to externalities (uncompensated social or environmental benefits or costs) (Abbott and Sumaila 2019, Franklin Associates 2014, Pilz, Brandt, and Fehringer 2010). For example, glass bottle substitutions would reduce the external cost of marine plastic waste. At the same time, replacing plastic beverage bottles with glass would increase private costs due to higher raw material prices and increased transportation costs due to the increase in weight. The increase in weight would also increase carbon emitted in transport, an external environmental cost.

This type of benefit-cost analysis that considers all values could be a powerful tool to reach sustainability goals. By including the value of ecosystem goods and services into the total economic value, reducing plastic waste can be used to preserve the oceanic natural capital and its services. Additionally, benefit-cost analyses can assist in reaching economic and welfare objectives. Currently, as negative externalized costs rise, the positive relationship between gross domestic product growth and welfare decreases, vanishes, or even becomes negative (Daly 2019). Importantly, environmental justice must be considered because the benefits and costs are not distributed equally—socially, geographically, or ecologically (see Box 1.3 for a more complete discussion of the unequal impacts of plastics). While internalizing externalities will increase direct costs, the benefit-cost analysis can assess whether something is economically wise (e.g., internal costs are lower than external costs). To make this assessment, natural capital must be valued.

While circular economic principles attempt to internalize externalities by significantly increasing the recycling of materials, lengthening product lifetimes, and primarily using renewable resources (Daly 2019,

World Economic Forum, Ellen MacArthur Foundation, and McKinsey & Company 2016), the net plastic production is projected to increase over time (Figure 2.6). The plastic industry expects growing populations and rising household incomes in much of the world to create new markets for the increased global plastic production capacity (CIEL 2018, UNEP 2014), which will ultimately result in increased plastic waste. However, it is unclear whether it is the consumers who want plastics or producers using plastic packaging as a cheap material to ensure their product's longevity and safety. Unless efforts are undertaken to more effectively manage the increased waste, an increase in environmental contamination by plastic waste will likely result (Borrelle et al. 2020). Given the important role of economics as a driver of both plastic production and consumption as well as recycling (Issifu, Deffor, and Sumaila 2021), the use of economic instruments to reduce plastic pollution is one of the levers available to both governments and private actors (Abbott and Sumaila 2019). Steps to internalize externalities of plastics, whether by adopting circular economy principles or by other means, can reduce plastic consumption, production, and waste streams.

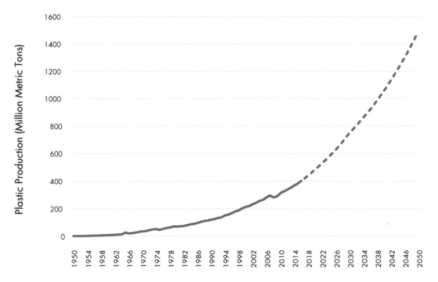

FIGURE 2.6 Global plastic production trend and projected growth. SOURCES: Data from 1950 to 2015 from Geyer, Jambeck, and Law (2017) supplemental material; projected numbers from Ellen MacArthur Foundation's annual industry growth estimations (World Economic Forum, Ellen MacArthur Foundation, and McKinsey & Company 2016)—2016–2020 has an annual 4.8% growth rate, 2021–2030 4.5%, and 2031–2050 3.5%. This does not include COVID-19 impacts.

CHAPTER SYNOPSIS

The United States plays a major role in global plastic production by producing, importing, and exporting plastic resins and plastic products. The vast majority of plastics are made from fossil sources. The economics of plastic production, which are linked closely with the fossil-based energy industry, have created barriers to adopting new "circular economy" concepts designed to conserve resources and reduce waste—from recycling to creating innovative reuse systems and developing new materials with end-of-life management as a primary design principle. In addition, it is important, but challenging, to communicate to consumers, policy makers, and others the nuances associated with material biodegradability and compostability, including appropriate use and management at end of life. While such new materials must play a role going forward, they are not a panacea. Furthermore, the natural environment should not be considered a large-scale, viable option for waste treatment.

As discussed in the next chapter, when plastic resins or products become plastic waste, those responsible for achieving an effective "end of life" for that plastic waste (reuse, recycle, dispose, compost)—from consumers and communities to local and state governments—face major systemic, economic, and policy barriers. Some of these barriers are "baked in" at the plastic production stage. There is a major economic and societal need and opportunity for the sectors involved in plastic production to formulate and design plastics and plastic products with a viable and safe end of life in mind. Some of this work is beginning or continuing, but an increase in scale is needed to meet the plastic waste challenges associated with current and forecasted plastic production and use.

PRIORITIZED KNOWLEDGE GAPS

A major data gap for plastic production is having transparent and accessible data on plastic production. As seen in this chapter, data on a wide array of plastic resin types could not be found and utilized. Without having access to plastic production data, it can be difficult to anticipate and react to production changes or to determine effective strategies to bolster plastic waste management.

FINDINGS AND CONCLUSION

Finding 1: Virgin plastic prices are artificially low due to fossil fuel subsidies; therefore, virgin plastics are more profitable to produce.

Finding 2: Approximately one-fifth (19%) of 2019 global plastic production occurred in North America, second to Asia. U.S. production of virgin plastics continues to increase, in part due to low costs of production for fossil-based feedstocks and rising production capacity.

Finding 3: The complex international system of plastic production, trade, and use complicates efforts to fully quantify the role of the United States in plastic production, export, import, use, and the country's contribution to plastic pollution.

Conclusion 1: Because the vast majority of plastics are carbon-carbon backbone polymers and have strong resistance to biodegradation, plastics accumulate in natural environments, including the ocean, as pervasive and persistent environmental contaminants.

3

Plastic Waste and Its Management

O nce produced, plastics are formed into a range of products that are used for a period of time. Some products, such as packaging, may have a very short use time while other more durable plastic products may remain in use for decades. There can be a short or long lag time between plastic production and its transformation into plastic waste. Plastic waste is created when, intentionally or unintentionally, plastics are taken out of use and enter a waste stream as part of a waste management process or are released into the environment.

This chapter first presents global estimates of plastic waste, followed by a detailed look into U.S. municipal solid waste (MSW) characterization, generation, and management. Other sources of U.S. plastic waste are explored. "Leaks" of plastic waste into the environment are discussed. Lastly, this chapter reviews the current regulatory framework of plastic waste management in the United States. Subsequent chapters identify transport, pathways, distribution, and fate of plastic waste that leak to the environment and ultimately to the ocean.

NATIONAL AND GLOBAL PLASTIC WASTE GENERATION

Plastic waste generation is directly related to the quantity of plastics produced and used. Understanding and estimating plastic waste generation can be challenging; there are a few different estimates from the past few years, which are summarized in Table 3.1. In terms of cumulative generation of plastic waste, Geyer, Jambeck, and Law (2017) estimate that from 1950 through 2015, 6.3 billion metric tons (BMT) of plastic waste were generated globally (Figure 3.1). In addition, Geyer, Jambeck, and Law (2017) estimated that in 2015, 302 million metric tons (MMT) of global plastic waste were generated. According to World Bank annual estimates, in 2016, the world generated 2.01 BMT of waste, of which 242 MMT was estimated to be plastic waste (Kaza et al. 2018). With cumulative quantities of plastic production projected to reach 34 BMT and plastic waste projected to reach 26 BMT by 2050, the total amount of plastics in the waste stream is projected to grow (Geyer, Jambeck, and Law 2017) (Figure 3.1).

Table 3.1 also indicates national estimates for U.S. plastic waste generation with estimates of 42 MMT in 2016 by Law et al. (2020) and 32 MMT in 2018 by the U.S. Environmental Protection Agency (U.S. EPA 2021c).

U.S. MANAGEMENT OF PLASTIC WASTE

Municipal Solid Waste

This chapter describes solid waste management and primarily focuses on MSW, what people throw away every day at home and on the go. It is typically measured in mass per person (per capita) generation rates. This chapter does not include intentional/permitted or unintentional land-based air, water (whether wastewater, stormwater, or other water), or

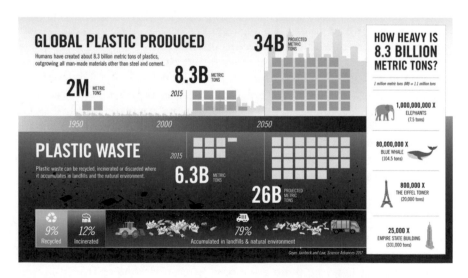

FIGURE 3.1 Global plastic production and waste generation infographic. SOURCE: Geyer, Jambeck, and Law (2017). Graphic credit: University of Georgia.

TABLE 3.1 Recent Estimates of Annual and Cumulative Generation of Plastic Waste in the United States and Globally

Data Source	Annual Plastic Waste Generation		Cumulative Waste Generation Since ~1950	
	USA	Global	USA	Global
U.S. EPA 2021c	32 MMT in 2018	–	[1,000] MMT	–
Law et al. 2020	42 MMT in 2016	–	–	–
Geyer, Jambeck, and Law 2017	–	302 MMT in 2015	–	6,300 MMT in 2015
Kaza et al. 2018		242 MMT in 2016		

NOTE: Square brackets indicate "on the order of" or "approximately." These estimates were completed by the committee using available data.

sludge (e.g., from wastewater treatment plants) discharges that may also contain plastics (usually smaller particles such as pre-production plastics or microplastics from clothing) unless they are disposed of as solid waste. It also does not apply to marine discharges (e.g., lost during shipping, lost or discarded fishing gear) unless recovered and deposited in a solid waste management system. Information on non-solid waste discharges and leakage is included in subsequent chapters.

Municipal Solid Waste Generation

The U.S. per person MSW generation rate ranges from 2.22–2.72 kg/day (4.9–6 lb/day) (EREF 2016, Powell and Chertow 2019, U.S. EPA 2021e). This is 2–8 times the waste generation rates of many other countries (Law et al. 2020). Figure 3.2 can be examined to see other countries' waste generation per capita. The United States generated about 321 MMT of waste in 2016, amounting to 16% of the world's waste (Kaza et al. 2018, Law et al. 2020). In 2016, the United States was the top generator of plastic waste (Law et al. 2020). This is despite containing 4.3% of the world's population (World Bank 2021) and being the third most populous country in the world.

In theory, managed solid waste in the United States should not contribute to ocean plastic waste because it is contained by treatment and/or conversion into other products (recycling, composting, incineration) or contained in an engineered landfill environment. In practice, plastic waste still "leaks" from managed waste systems when blowing out of trash cans,

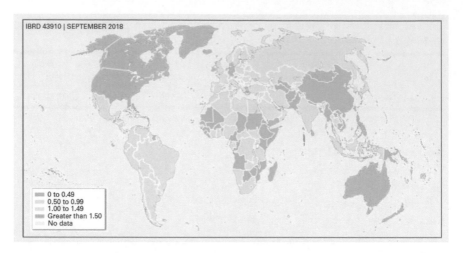

FIGURE 3.2 Waste generation per capita, illustrated in kilograms. SOURCE: Kaza et al. (2018).

trucks, and other managed scenarios. Waste not put into the management system, whether intentionally or unintentionally through actions such as illegal dumping and littering, is considered unregulated and illegal waste in the United States.

Data on MSW are compiled by U.S. EPA through a materials flow analysis method. The quantities are estimations based on production, along with lifetimes for various products and sectors to estimate the quantity of waste generated in each sector and for particular products. Data are also measured by other industry and academic groups, states, and even cities to inform local waste management. The management of MSW typically takes place at the city or county level in the United States, and nearly every household is provided with a method to formally manage their waste. Other waste streams in the United States that may contain plastics also are described in this chapter, although little is known about their contribution to ocean plastic waste.

Municipal Solid Waste Characterization

U.S. EPA's Sustainable Materials Facts and Figures report, which calculates estimates as far back as 1960 and has been published periodically for more than 20 years, focuses on MSW. According to U.S. EPA, the MSW items include "packaging, food, grass clippings, sofas, computers, tires and refrigerators." However, U.S. EPA does not include in its analysis any materials disposed of in non-hazardous landfills that are not generally considered MSW such as construction and demolition debris, municipal wastewater treatment sludges, and non-hazardous industrial waste, some of which may be composed of plastics.

According to U.S. EPA, the generation of waste is the

> weight of materials and products as they enter the waste management system from residential, commercial, and institutional sources and before recycling, composting, combustion or landfilling take place. Pre-consumer (industrial) scrap is not included in the waste generation estimate. Source reduction activities, such as backyard composting of yard trimmings, take place ahead of generation.

U.S. EPA's materials flow methodology does not consider any "mismanagement" of waste within the United States, such as illegal dumping or littering.

The U.S. EPA MSW characterization describes waste both by material type—paper, plastics, metal, glass, etc.—and by-products, which are separated into durable goods (typically stay in use more than 3 years), nondurable goods (stay in use less than 3 years), and containers and packaging (typically enter the waste stream the same year they are purchased). Examples of durable goods include appliances, furniture, casings of lead-acid batteries, and other products. Examples of nondurable goods include disposable diapers, trash bags, cups, utensils, medical devices, and household items such as shower curtains. U.S. EPA does not include plastics in transportation products, other than lead-acid batteries, in its management analysis (U.S. EPA 2021e).

U.S. EPA estimated that 12.2% of MSW (by mass) was plastics (32.4 MMT) in 2018. However, the estimate for annual generation of plastic solid waste has been as high as 42 MMT when using waste generation rates derived from waste disposal data from MSW management facilities (Law et al. 2020). Plastics are the third-highest percentage of material (by mass) in MSW after paper and food waste, and are slightly higher than yard waste (Figure 3.3).

The steep increase in plastic production described in the previous chapter has been mirrored by an increase in the percent of plastics in U.S. MSW (by mass)—from 0.4% in 1960 to 12.2% in 2018, with a peak of 13.2% in 2017 (U.S. EPA 2020a). The mass of plastic waste generated has been increasing in the United States since 1960, with the fastest increase occurring from 1980 to 2000 (Figure 3.4).

Municipal Solid Waste Collection

Residential waste is a category of MSW. MSW is broader and includes waste from single-family homes to multi-family housing and waste from commercial and institutional locations, such as businesses, schools, and hospitals. Generally, single-use plastics used in the home and packaging for

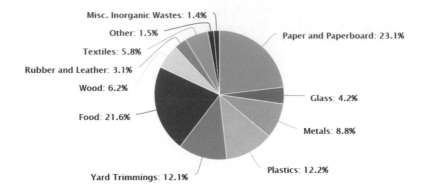

FIGURE 3.3 Municipal solid waste generation categorization by mass in the United States for 2018. SOURCE: U.S. EPA (2021e).

any packed food items will end up in the residential waste stream, as will longer-lived durable goods, when disposed of. In the United States, the residential waste and recycle stream usually is picked up at people's homes by the local community (paid through either fees or taxes) or a private hauler (hired by the resident), or the resident takes the waste to a transfer station or

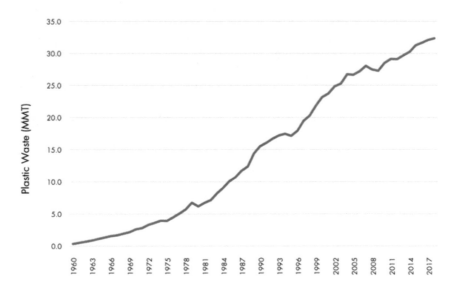

FIGURE 3.4 U.S. annual plastic waste generation from 1960 to 2018 in million metric tons. SOURCE: U.S. EPA (2020a).

directly to a management facility (e.g., landfill or recycling facilities called material recovery facilities [MRFs]). Plastic waste generation at the residential level is not measured or monitored directly. Community members typically do not know how much or what kind of waste they generate. Residential waste and mass of items collected for recycling are recorded at the community level through landfill or MRF disposal. Garbage truck weight is measured at the landfill scale houses to calculate tipping fees (e.g., a fee to pay for waste disposal). Outgoing trucks of baled materials (e.g., bales of plastics, such as polyethylene terephthalate [PET] or mixed plastics) that are shipped to processing facilities for recycling are also weighed.

Since solid waste is typically measured in mass (e.g., for solid waste audits, "tipping" fees at disposal facilities), but plastic bulk density is low, it weighs very little for how much space it takes up if uncompacted. The bulk density (the weight of the waste divided by the volume it occupies, including the space between waste items) of uncompacted mixed plastics is approximately 121 lb/yd^3 (72 kg/m^3). For example, trash may look like it is composed mostly of plastics because film plastics spread out and look large owing to their surface area, and empty plastic containers still take up the space that held the product.

Waste collection methods are often determined by population density. For low population densities, curbside collection may not be economically feasible and residents may be required to take their own waste to a transfer station for drop-off, which puts an extra burden on residents. Rural areas not served by curbside collection may manage more MSW, including plastics, "at home" through open burning and dumping privately/illegally (Tunnell 2008). In Virginia, for example, open burning is still allowed if there is no regular trash collection.[1] With population density as a driver for waste generation, higher density areas such as urban and suburban areas generate more plastic waste per unit area than rural areas; however, urban areas have more developed waste management infrastructure (e.g., more curbside collection and recycling) than rural areas. This pattern occurs globally as well as in the United States (Schuyler et al. 2021, Youngblood et al. In Review).

Although plastic waste quantities generated in urban and rural areas differ and the proportion of plastic waste not collected or captured by waste management systems varies, both are sources of ocean plastic waste (see subsequent chapters). Regardless of population density or land use, coastal areas have greater connectivity to the ocean, placing any

---

[1]Code of Virginia § 10.1-1308; Clean Air Act; §§ 110, 111, 123, 129, 171, 172, and 182; 40 CFR Parts 51 and 60. "Open burning is permitted for the on-site destruction of household waste by homeowners or tenants, provided that no regularly scheduled collection service for such refuse is available at the adjacent street or public road."

uncollected plastic waste from urban, suburban, rural, recreational, industrial, or other human activities at a higher risk of ending up in the ocean. Coastal areas might be subject to greater efforts to reduce, collect, and divert plastic waste sources, but inland areas, especially along waterways, should be managed to reduce plastic wastes moving toward the ocean.

Municipal Solid Waste Management

In 2018, to manage MSW, the United States landfilled 50%, recycled 24%, composted 8.5%, and combusted 12% of all MSW (U.S. EPA 2021e). Of plastics in MSW, 75.6% were landfilled (comprising 18.5% of all landfilled materials, by mass), 8.7% were recycled, and 15.8% were combusted with energy recovery. While both recycling and combustion capacity expanded in the 1980s and 1990s, these percentages have remained relatively consistent over the past 15 years (Figure 3.5).

Decisions about how waste, including plastic waste, is managed are made by state and local governments and other groups, who bear the growing costs and challenges of managing increasing amounts of waste. Plastic products disposed as waste (reported by U.S. EPA in durable goods, nondurable goods, and containers and packaging categories) consist of a wide variety of plastic polymers containing mixtures of chemical

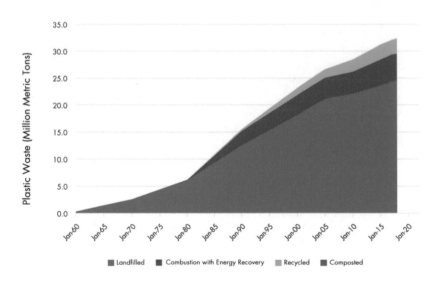

FIGURE 3.5 U.S. plastic waste management of municipal solid waste from 1960 to 2018 in million metric tons (MMT) per year. Composted levels are at zero during this period. SOURCE: U.S. EPA (2020a).

additives that allow for an array of properties (Deanin 1975). Thus, the composition of plastics in MSW is incredibly diverse, which creates challenges in waste management systems, especially when sorting materials for appropriate recycling or composting.

Landfilling

Since the Resource Conservation and Recovery Act (RCRA) passed in 1976, landfills are lined with composite liners to protect the soil and groundwater (e.g., geomembrane and 2 feet of compacted clay), and the liquid that permeates and seeps through the landfill waste is collected and removed. Landfills are sloped to one side with a drainage layer (e.g., sand) so the liquid can quickly run off the liner, collect, and then be pumped out of the landfill. Trucks deposit waste onto the working face of the landfill and bulldozers move the waste. Compactors compress the waste so the landfill is as dense as possible. Once the landfill has reached its fill height, gas wells are installed throughout the landfill to collect released gases (i.e., methane, carbon dioxide, nitrogen, and other trace gases). The landfill is then capped with an impermeable layer, which is similar to the bottom layer. Sometimes soil and grass are placed on top of the landfill. After the landfill is closed, it requires at least 30 years of monitoring.

None of the highest-production plastics (PET, high-density polyethylene [HDPE], polyvinyl chloride [PVC], low-density polyethylene, polyethylene [PE], polystyrene [PS]) biodegrade in a landfill, and they are considered contamination in compost. Since plastic products also contain an array of additives (Deanin 1975), this diversity of plastic waste can challenge recovery and recycling. In addition, plastics can be mixed with food waste, most of which goes to landfills (only 6.3% of food waste is composted, as compared with 69.4% of yard waste, which is restricted from landfills).

With the vast majority (76%) of managed plastic waste disposed of in landfills, there are opportunities to reduce this amount and conserve non-renewable resources, increase energy efficiency, and provide economic and environmental benefits through effective source reduction, recycling, and composting. These options are in line with U.S. policy to prevent and reduce pollution at the source whenever feasible (Pollution Prevention Act). These principles are expressed in the RCRA, where the order of preference in managing materials is source reduction, reuse, recycling, and disposal.

Recycling

The statistics reported by U.S. EPA on plastic recycling reflect the amount of plastic waste collected for reprocessing into a secondary raw material, primarily by mechanical recycling. Mechanical recycling requires waste items to first be sorted according to primary material type (polymer resin type), indicated on many household products by the numbered resin identification code ("chasing arrows" symbol). Products might be further sorted according to color, size, or density before being washed of residues or contaminants, then shredded or chopped into smaller particles that can be remelted and formed into a reprocessed material (Ragaert, Delva, and Van Geem 2017).

The increasing diversity and complexity of material and product types present major challenges to recycling, especially when waste is collected in "single-stream" recycling programs, which require mechanical and manual separation at MRFs. Contamination of individual plastic items by food or product residues, and of entire loads by items that are not recyclable (often by people "wish-cycling," who place items in recycling collection in hopes they might be recycled), increases the difficulty and cost of separation (Damgacioglu et al. 2020). Furthermore, because plastics degrade throughout their life cycle and during reprocessing, recycled materials are frequently used in "downcycling" applications that do not require the same material quality standards as food-grade applications, for example (Ragaert, Delva, and Van Geem 2017). For these reasons and others, such as the low cost of primary (usually fossil) feedstocks used to make virgin plastics and fluctuating market demand for recycled materials, the economics of recycling can be extremely challenging (Rogoff and Ross 2016). Further details on where plastic scrap can be exported are illustrated in Box 3.1.

A suite of chemical processes, many of which are under development, that aim to break plastic waste down into chemical constituents, which may include the monomer building blocks of the original plastic (total depolymerization) or other intermediates (partial depolymerization), are broadly referred to as "chemical recycling" or "advanced recycling". A major goal of chemical recycling is to produce secondary materials of the same or higher quality than the initial plastic waste itself ("upcycling"), ideally striving for many cycles of polymerization and depolymerization to maximize resource use (Coates and Getzler 2020). Presently, the only forms of chemical recycling utilized in the United States (and only at small scale) are energy-intensive pyrolysis and gasification processes, whose primary products are fuel and other chemical products rather than secondary polymers (Ragaert, Delva, and Van Geem 2017). Priority research opportunities have been identified to inform federal investment in research into new materials, together with the chemical processes to upcycle these materials once they become waste, to move toward a more circular life cycle for plastics (Britt et al. 2019).

Challenges include incompatibility of different plastic types and large differences in processing requirements (Closed Loop Partners 2020, Hopewell, Dvorak, and Kosior 2009, OECD 2018). Addressing these barriers to plastic recycling can produce co-benefits, including improving energy

BOX 3.1
Management Through Import and Export of Plastic Scrap

Some of the plastic materials sent to material recovery facilities in the United States are exported to other countries after processing. Before the import restrictions initially implemented by China at the end of 2017 (resulting in a relative import ban), the United States exported half of its plastic waste intended for recycling to China (Brooks, Wang, and Jambeck 2018). After 2018, plastic scrap previously destined for China was either re-routed to other countries (e.g., Cambodia, India, Indonesia, Malaysia, Pakistan, Vietnam, Thailand, and Turkey) or placed in domestic landfills (INTERPOL 2020). U.S. plastic scrap exports decreased by 37.4% in the first quarter of 2018, largely due to the 92.4% decline in plastic scrap exports to China (Mongelluzzo 2018). In the same time period, U.S. waste exported to Malaysia increased by 330%, to Thailand by 300%, to Vietnam by 277%, to Indonesia by 191%, and to India by 165% (INTERPOL 2020). In 2018, other Asian countries (e.g., Indonesia, Thailand, Malaysia, Vietnam, Taiwan, and India) started to regulate, and sometimes ban, plastic waste imports due to waste surpluses and illegally exported wastes (e.g., hazardous waste mixed in with plastic scrap) (INTERPOL 2020, Staub 2021, Upadhyaya 2019). In 2020, the United States' top six trade partners (Canada, Malaysia, Hong Kong, Mexico, Vietnam, and Indonesia) accounted for 75% of U.S. exports of plastic scrap (Brooks 2021).

Export destinations of U.S. plastic waste can be a source of plastics in the ocean. Recent amendments to the Basel Convention on the Control of Transboundary Movements of Hazardous Wastes and Their Disposal placed new controls on exports of plastic waste. However, the United States is not a signatory and is therefore not subject to the stricter guidelines of plastic exports. As such, U.S. plastic waste exports have continued, though greatly decreased as described above. In addition, U.S. exports will be affected by decisions of the receiving countries that are parties to the Convention (U.S. EPA 2021d). In the absence of the Basel Convention, the United States could continue to record and document exports by the U.S. Trade Association and UN Comtrade.

efficiency, environmental performance, and process efficiency, while creating economic opportunities for new products (U.S. Department of Energy 2021). A variety of prizes or challenge competitions have been designed to stimulate innovation in overcoming the barriers associated with plastic recycling or to minimize reliance upon these difficult-to-manage materials (e.g., Department of Energy Plastics Innovation Challenge, New Plastics Economy Innovation Prize, the REMADE Institute, or the Bio-Optimized Technologies to keep Thermoplastics out of Landfills and the Environment [BOTTLE] Consortium), and some of these efforts have already produced results (Rorrer, Beckham, and Roman-Leshkov 2021, Shi et al. 2021).

Composting

High-production plastics such as PE, polypropylene, PS, and PVC are strongly resistant to biodegradation in any environment, due to the strength of the carbon-carbon bond that constitutes the polymer backbone. Therefore, managed composting is not a suitable management strategy for the vast majority of today's plastic waste, which would be contaminants in composting environments. A variety of certified compostable plastics (with ester backbones) have been developed to completely biodegrade (defined by complete metabolism by microorganisms in a specified time period) in managed composting facilities that maintain the specific environmental conditions required for material breakdown. However, the benefits of these products are lost if they are not collected and transported to managed composting facilities. In most regions of the United States, such facilities are not available. Even if there are nearby facilities, the consumer must recognize the item as compostable and place it in the correct collection bin, rather than in regular trash or in recycling collection, where it would contaminate the recycling stream (Law and Narayan 2022). Thus, the benefits of compostable plastics can only be realized if sizeable investments in composting infrastructure and consumer education occur.

Management of Plastic Containers and Packaging

Plastic containers and packaging comprise the largest fraction of the plastic waste stream (41%) and enter the waste stream most quickly after production in the year they are produced. Products in this category also commonly leak from the waste management system (see subsequent section on leakage). U.S. EPA defines plastic packaging as bags, sacks, and wraps; other packaging; PET bottles and jars; HDPE natural bottles; and other containers. It does not include single-service plates, cups, and trash bags, all of which are classified as nondurable goods. Plastic containers and packaging were the highest category within plastic materials in 2018 with an estimated 13.2 MMT generated, or approximately 5.0% of total MSW generation (U.S. EPA 2021c). In 2018, 1.8 MMT (13.6%) of plastic containers and packaging materials were recycled. However, this was lower than the quantity combusted with energy recovery, 16.9% (2.2 MMT), while the remainder (more than 69%) was landfilled (Figure 3.6). The two items most commonly recycled were PET bottles and jars at 29.1% (of total PET bottle waste generation) and HDPE natural bottles (e.g., milk and water bottles) at 29.3% (of total HDPE natural bottle generation). The higher rates of recycling are reflective of the product mass, with containers heavier than film plastics, and their more uniform design characteristics (monochromatic and with fewer additives), which makes these products easier to recycle and the recycled material more valuable.

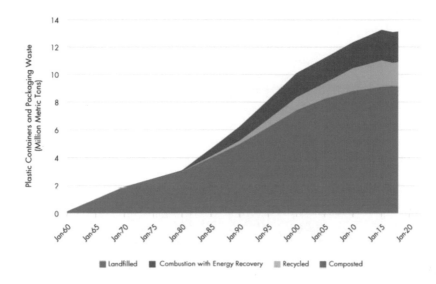

FIGURE 3.6 U.S. Environmental Protection Agency data on plastic containers and packaging waste management. Composted levels are at zero during this period. SOURCE: U.S. EPA (2021c).

Management by Designing for End of Life

The approach of designing products for end of life is embedded in the U.S. EPA's Sustainable Materials Strategy and related programs (U.S. EPA 2015). However, there are many barriers, including a substantial mismatch between the materials that are created and the ability of the waste management system to accept and transform these materials into a second use or beneficial product (U.S. GAO 2020), such as being effectively recyclable or biodegradable.

Part of the solution to this mismatch is to adopt an integrated, life-cycle perspective (Walls and Palmer 2001) in the design of plastic products, especially single-use products, that explicitly accounts for direct and indirect costs associated with the product's end-of-life disposal. This perspective would reduce the social cost of plastic disposal and waste leakage by pushing producers to design and use more easily biodegradable and recyclable/reusable materials, and by enabling consumers to choose products that permit low-impact disposal (Abbott and Sumaila 2019). Green Engineering principles (American Chemical Society 2021), if followed during material development and product design, can reduce the externalities associated with plastics. Circular economy concepts, designed to promote "a regenerative system in which resource input and waste, emission, and energy leakage are minimized by slowing, closing, and narrowing material and energy loops thanks to long-lasting design, maintenance, repair, reuse, remanufacturing, refurbishing, and recycling" (Geissdoerfer et al. 2017), may be helpful as well.

Developing alternative materials or other product delivery systems can spark innovation and economic growth in the United States. There are several voluntary corporate commitments to change materials, use more recycled materials, and increase material circularity, so materials and infrastructure development to meet those demands are needed.[2] Efforts could include sustainable packaging associations (precompetitive collaborations) to develop alternative materials and agree on more homogenized packaging designs for end of life, packaging with more value (e.g., single, homogenous materials; design for recycling/end of life), and designing out problematic items/materials (e.g., certain colors, smaller caps/lids). For composting to be a part of an integrated management approach, there is a need for both biodegradable materials and further development and expansion of composting infrastructure in the United States. For a more detailed approach to materials design, please see the recent article by Law and Narayan (2022).

Municipal Solid Waste Management Disparities and Environmental Justice

U.S. EPA defines environmental justice as "the fair treatment and meaningful involvement of all people regardless of race, color, national origin, or income with respect to the development, implementation and enforcement of environmental laws, regulations and policies." (U.S. EPA 2021b). Environmental justice is one of the top priorities of the current U.S. EPA Administrator, Michael S. Regan (U.S. EPA 2021b). Impacts to vulnerable populations occur all along the life cycle of plastics, starting from the extraction of oil and natural gas as feedstocks of plastic production and including the production of plastic resins at refining and chemical processing facilities, the use of plastics from smaller or limited packaging choices, and management and leakage of plastic waste to the environment (CIEL 2019, UNEP 2021b).

Environmental justice efforts around waste began in the United States with communities (e.g., in Houston, Texas and Warren County, North Carolina) fighting landfills and hazardous waste management facilities in areas populated predominantly by African Americans (Bullard 1990, McGurty 2000). These impacts and concerns continued for years, with

---

[2]See https://usplasticspact.org/.

research similar to that done on hazardous waste landfills conducted on U.S. non-hazardous solid waste landfills in the contiguous 48 states finding that these landfills are also more likely to be located in counties with higher percentages of poverty and people of color (Cannon 2020). More recently in Houston and Dallas, Texas, studies show people of color are concentrated in neighborhoods closer to MSW landfill facilities where housing prices and median incomes are lower than those just 2 or 3 miles away (Erogunaiye 2019). This research also showed that the magnitude of disparity within 1–3 miles of a landfill had increased over the 15-year period from 2000 to 2015 (Erogunaiye 2019). Additionally, MSW incinerators are disproportionately located in communities with at least 25% people of color and/or impoverished people (Tishman Environment and Design Center 2019). Burning plastics releases toxic chemical pollutants, such as dioxins and furans (Verma et al. 2016), which can have serious health implications for community members (Tishman Environment and Design Center 2019, Verma et al. 2016, and see Box 1.3 for more information on health impacts).

U.S. EPA, in line with the Biden-Harris Administration's directive to all federal agencies to "embed equity into their programs and services to ensure the consistent and systematic fair, just, and impartial treatment of all individuals," announced in April 2021 that it was taking steps to address environmental justice across the agency. These steps include strengthening enforcement of violations, incorporating environmental justice across all its work, improving "early and more frequent engagement with pollution-burdened and underserved communities" and tribal officials, and considering and prioritizing "direct and indirect benefits to underserved communities in the development of requests for grant applications and in making grant award decisions as allowed by law" (U.S. EPA 2021b).

Municipal Solid Waste COVID-19 Impacts

The global COVID-19 pandemic has had extensive impacts on the generation and characterization of MSW in the United States. Within 1 week of various city, state, or national mandates for public areas to use and wear personal protective equipment, such as masks, these items were reported as litter through the Marine Debris Tracker mobile app and to programs of the Ocean Conservancy (Ammendolia et al. 2021, Marine Debris Tracker 2020, Ocean Conservancy 2021). In addition, waste collection companies reported decreases in commercial waste collection because people were not commuting to the office or conducting activities outside of home (*Waste Advantage Magazine* 2020). For the same reasons, residential waste increased by 5–35%, increasing logistical and economic strain on haulers and communities trying to manage MSW (Dzhanova 2020, Redling 2021).

Other Types of Plastic Waste (Non-MSW)

While some waste categories are included in the measurement of MSW, some other sources of plastic waste are identified below. Only some are measured or monitored under existing federal environmental law. The most consistent and well-documented information on U.S. plastic waste comes from data on management of solid waste under RCRA or documentation of waste recovered from or measured in the environment (see Chapters 4 and 5). Because many leakage estimates rely only on MSW data, they are likely conservative estimates. Aside from the National Oceanic and Atmospheric Administration's (NOAA's) Marine Debris Monitoring and Assessment Program (Chapter 6), no federal monitoring programs document or monitor the amount of plastic waste contained in air or water discharges, though state and local governments have conducted specific monitoring studies, sometimes with federal support or assistance.

Construction and Demolition Debris

Starting in 2018, U.S. EPA included construction and demolition debris as a separate section outside of the MSW waste generation in its Sustainable Materials Facts and Figures report (U.S. EPA 2021e). In general, construction and demolition debris materials are durable goods and do not enter the waste stream quickly. However, they are sometimes illegally dumped or managed at unregulated construction sites or abandoned lots and it is unknown what quantity may be entering the ocean.[3] Construction and demolition debris is also generated in catastrophic events (e.g., hurricanes, tsunamis, floods, etc.), which can generate debris, including plastics, that enters waterways and the ocean. The most prominent example of this occurred when the Tohoku Tsunami hit Japan. Of the 5 MMT of debris generated, 1.5 MMT floated and portions subsequently were transported to the shores of the United States (Murray, Maximenko, and Lippiatt 2018). It is currently unknown how much plastic waste may enter the ocean in U.S. waters from catastrophic events, such as floods.

Industrial

Industrial waste is any waste (including plastics) generated by manufacturing or industrial processes. As solid waste, it can be classified under RCRA as either hazardous or non-hazardous solid waste, and governed by assigned management requirements (see Appendix C: Legal Framework for more information). Industrial waste can include plastic pellets, also referred to as nurdles.

---

[3]J. Jambeck, University of Georgia, personal communication, 2021.

Industrial waste can also include sludge and liquid waste from industrial facilities regulated and permitted under other statutes, such as the Clean Water Act (U.S. EPA 2021g); however, the Clean Water Act does not identify plastics as a pollutant for discharge monitoring or limits (Appendix C). However, some chemicals used in plastics (and many other industrial applications) may be separately monitored or regulated. Under the Pollution Prevention Act, which promotes pollution prevention and production, U.S. EPA collects and publicly shares data on industrial facility releases of certain harmful chemicals (including unregulated chemicals) that it lists on the Toxics Release Inventory (TRI) (U.S. EPA 2021i). The TRI does not include plastics but does include several chemicals used in the manufacture of plastics (Wiesinger, Wang, and Hellweg 2021).

Plastic Waste in Wastewater and Stormwater

Some plastic waste enters wastewater infrastructure in sewage, sometimes combined with stormwater. Nearly all large plastic items entering sewers and arriving at wastewater treatment plants are removed by bar screens before treatment through biological and chemical processes. Most microplastics remain in the post-treatment sludge (managed typically through landfilling or land application) with a smaller amount discharged in treated wastewater, mostly as small fibers and fiber fragments (Carr, Liu, and Tesoro 2016). No federally mandated monitoring of plastic waste occurs at wastewater treatment plants. A 2021 U.S. EPA multisector stormwater general permit has been challenged in court for not sufficiently addressing plastic pollution from pre-production plastic pellets, flakes, and powders (Center for Biological Diversity 2021, U.S. EPA 2021g).

Transportation Infrastructure

Transportation systems are sources of plastic waste in the environment, including plastics shed from the operation of transportation systems (e.g., from tires, paints, brake linings), litter from passengers (considered MSW) and cargo, and litter from transportation systems themselves (e.g., plastics and chemicals from road paint and asphalts). Transportation systems also tend to be sources of plastics to stormwater and other drainage systems that transport plastic wastes to local waterways and as far as the ocean, with tire particles being a major source of microplastics (Werbowski et al. 2021), as described in Chapter 4. Some industrial plastics from transportation systems appear to have special forms of toxicity. For example, a tire-rubber-derived chemical called 6PPD-quinone (also known as (N-(1,3-dimethylbutyl)-N'-phenyl-p-phenylenediamine quinone)) has been identified as a cause of mortality for salmon in the

U.S. Pacific Northwest (Tian et al. 2021). Nonpoint source runoff from highways is subject to management guidance under the U.S. EPA Clean Water Act programs, as well as in coastal and Great Lakes areas through a joint program with NOAA under the Coastal Zone Management Act (U.S. EPA 2021f). However, current federal law does not require monitoring of the sources of macroplastics or microplastics in transportation systems (Appendix C).

Marine Activities

The disposal of plastic waste from vessels and at-sea platforms into the ocean is prohibited by the 1988 international maritime regulations (MARPOL Annex V). The United States is a signatory to MARPOL Annex V (an optional, non-mandatory annex of MARPOL), which has been incorporated into U.S. law via the Act to Prevent Pollution from Ships (33 USC § 1901 and 33 CFR Part 151). However, enforcement of MARPOL Annex V is challenging and compliance is difficult to assess. In addition, accidental loss of plastic waste at sea occurs, such as from abandoned vessels, lost ships and cargo, and release of plastic products or plastic "nurdles" from shipping containers. Some of these losses are recognized at the state legislative level, such as abandoned vessels, which are subjects of public concern, but are not well quantified in the United States or U.S. waters.

One type of maritime-generated ocean plastic waste is abandoned, lost, or otherwise discarded fishing gear (ALDFG). No robust estimates of the total amount of ALDFG generated worldwide or by U.S. domestic fisheries are available (Richardson et al. 2021), though a recent global meta-analysis indicates 5–30% of fishing gear is lost annually worldwide depending on gear type (Richardson, Hardesty, and Wilcox 2019). Industrial trawl, purse-seine, and pelagic longline fisheries are estimated to lose a median of 48.4 kt (95% confidence interval: 28.4 to 99.5 kt) of gear each year during normal fishing operations, but this estimate does not include abandoned or discarded gear; other gear known to become derelict such as pots and traps, poles and lines, and driftnets/gillnets; or gear from nearshore and small-scale fisheries (Kuczenski et al. 2022). The role of illegal, unreported, and unregulated fisheries in the generation of ALDFG, or other plastic waste, is also unknown. Lastly, ALDFG resulting from U.S. recreational or subsistence fishing activities is also a source of ocean plastics that is little quantified or understood. There is also growing attention to the contribution of aquaculture activities to plastic waste at a global scale (Sandra et al. 2020), but U.S. contributions have not been assessed. A full description of the types of ALDFG generated in the United States or resulting from U.S.-based fisheries or aquaculture is beyond the scope of this report.

U.S. PLASTIC WASTE LEAKAGE

Quantities (Mass)

"Managed" plastic waste is contained by treatment and/or conversion into other products (recycling, composting, incineration) or contained in an engineered landfill. If not effectively "managed" in these ways it may have intentionally or unintentionally "leaked" into the environment. Plastic waste not making it into (e.g., illegal dumping, litter) or leaking out of (e.g., blowing litter or unregulated leaking or discharge) our management systems is categorized as "mismanaged" plastic waste. Figure 3.7 represents ways waste may leak, even from a solid waste management system reaching 100% of the population. Once in the environment, wastes are more difficult to recover for later treatment or disposal.

Because U.S. EPA data on MSW does not quantify mismanaged solid waste that leaks into the environment, researchers have developed approaches to derive such estimates, drawing on U.S. EPA-reported data and other data sources. Law et al. (2020) quantified the U.S. contribution of mismanaged plastic waste to the environment as 1.13–2.24 MMT in 2016. Mismanaged waste included a model estimate for litter, illegal dumping, and estimates of exported plastics collected for recycling that were inadequately managed in the importing country. Litter—solid waste that is intentionally or unintentionally disposed of into the environment despite the availability of waste management infrastructure—was coarsely estimated as 2% of plastic solid waste generation (owing to a lack of mass-based estimates of litter rates). For 2016, the quantity of plastic litter estimated annually in the United States was 0.84 MMT (Law et al. 2020). Law et al. (2020) estimated that 0.14–0.41 MMT of plastics were illegally dumped (i.e., disposed of in an unpermitted area) annually,

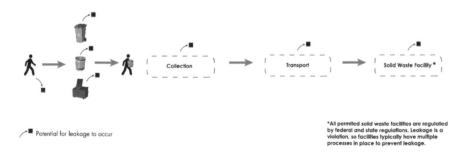

Potential for leakage to occur

*All permitted solid waste facilities are regulated by federal and state regulations. Leakage is a violation, so facilities typically have multiple processes in place to prevent leakage.

FIGURE 3.7 Points of plastic leakage for municipal solid waste in the United States. Black box with red outline denotes leakage potential.

despite the availability of waste management infrastructure. This estimate comes from assessment of illegal dumping in three U.S. cities (San Jose, California; Sacramento, California; and Columbus, Ohio).

The final component of mismanaged solid waste in the Law et al. (2020) analysis is exported plastic scrap collected for recycling that is inadequately managed in the importing country (see Box 3.1). Law et al. (2020) estimated that in 2016, 0.15–0.99 MMT of plastics exported by the United States in plastic scrap and paper scrap (in which plastics are included as contaminants) bales were disposed of during processing and likely entered the environment in the importing country (Law et al. 2020). The total quantity of plastic solid waste from the United States entering the environment in 2016 was estimated to be 1.13–2.24 MMT. Comparing mismanaged plastic waste from other countries, Law et al. (2020) concluded that the United States was the 3rd to 12th largest contributor of plastic waste into the coastal environment with 0.51–1.45 MMT in 2016.

High-Leakage Items

Similar to the waste management system categorizing the waste stream by material and products, varying plastic products and materials leak from the solid waste management system in different proportions evidenced by what does, and does not, end up in our environment. Litter surveys and community science efforts (at large scales, see Chapter 6) have shown that while plastics make up a large percentage (70–80%, see Table 3.2) of what is found in the environment as litter, the majority of plastic items are single-use, including packaging, as well as tobacco-related (e.g., cigarette filters, product packaging, and e-cigarette cartridges) (Public Health Law Center 2020) and unidentified fragments from larger items. These large-scale surveys generally do not include the documenting of microplastic or pre-production resin pellets at a more local level (Tunnell et al. 2020).

While historically marine litter studies and land-based work have not always been consistent in terms of methods used (Browne et al. 2015), there has been consistent, even if opportunistic, data collection through a few community science-based initiatives. These include the International Coastal Cleanup, which has been collecting data annually for more than 35 years; NOAA's Marine Debris Monitoring and Assessment Project initiative; and opportunistic data from the mobile app Marine Debris Tracker (initially funded by NOAA) as well as a scientifically designed targeted data collection event in the Mississippi River corridor in 2021 (Youngblood, Finder, and Jambeck 2021). For more information about these programs, please see Chapter 6 on Tracking and Monitoring.

TABLE 3.2 Top 10 Items Tallied from Each Data Set Compilation

Data Set	Date Range (*n* = number of litter items counted)	Top 10 in Rank Order
Ocean Conservancy's International Coastal Cleanup (USA only)	2015–July 2021 (*n* = 18,565,446), 82% plastic waste	Cigarette butts, food wrappers, plastic bottle caps, plastic beverage bottles, straws, stirrers, other trash, beverage cans, plastic grocery bags, glass beverage bottles, metal bottle caps, plastic lids
MDMAP Accumulation of items 2.5–30 cm	2009–2021 (*n* = 895,417), 84% plastic waste	Hard plastic fragments, foamed plastic fragments, plastic rope/net, bottle/container caps, filmed plastic fragments, plastic other, cigarettes, plastic beverage bottles, food wrappers
MDMAP Accumulation of items 30 cm or larger	2009–2021 (*n* = 5,561), 58% plastic waste	Lumber/building material, hard plastics, plastic rope/net, other plastics, cloth/fabric, foam plastics, film plastics, other metal, buoys and floats, other processed lumber, plastic bags
MDMAP 2.5 cm + standing stock and using MDMAP 2.0 protocol	2009–2021 (*n* = 71,306), 86% plastic waste	Hard plastic fragments, foamed plastic fragments, plastic bottle or container caps, plastic fragments film, plastic food wrappers, other plastics, cigarettes, plastic rope or net pieces, processed lumber–building material, plastic beverage bottles, processed lumber–paper and cardboard
Marine Debris Tracker (USA only)	2011–July 2021 (*n* = 2,333,337), 71% plastic waste	Plastic or foam fragments, cigarettes/cigars, plastic food wrappers, plastic caps or lids, other (trash), plastic bottle, plastic bags, paper and cardboard, aluminum or tin cans, foam or plastic cups or plates, straws
Mississippi River Plastic Pollution Initiative (MRPPI)	March 15–April 25, 2021 (*n* = 75,184), 74% plastic waste	Cigarette butts, food wrappers, plastic beverage bottles, foam fragments, aluminum cans, hard plastic fragments, plastic bags, plastic/foam cups, paper and cardboard, film fragments. Note: PPE was 1–2% of all litter found

NOTE: If an item labeled "Other" was in top 10, the 11th ranking item was also included since "Other" can include a wide array of items. MDMAP = Marine Debris Monitoring and Assessment Project, PPE = personal protective equipment.

The Cost of Leakage

While the drivers for leakage of plastics into the environment are complex and varied (see previous section), the cost and burden are borne by communities, especially residents. The United States spends roughly $11.5 billion on cleanup from trash leakage into the environment (Keep America Beautiful Inc. 2010). States, cities, and counties together spend at least $1.3 billion. Cleanup is often a hidden cost within employee salaries or other projects, which makes it difficult to determine the actual cost to local governments. For example, the Georgia Department of Transportation spends more than $10 million on annual labor and equipment costs necessary for picking up and disposing of trash from state roadways (GDOT 2020). CalTrans costs grew from $65 million in 2016–2017 to $102 million in 2018–2019 to keep trash off of transportation areas (CalTrans 2020).

CURRENT REGULATORY FRAMEWORK FOR U.S. MANAGEMENT OF PLASTIC WASTE

Starting in the 1970s, the United States created several legal frameworks designed to control and prevent the release of harmful, toxic, or hazardous substances, as well as manage transportation, treatment, and disposal of specific wastes. This body of law applies to many materials originally created for societal benefit that were later found to be harmful to human or environmental health, such as polychlorinated biphenyls or chlorofluorocarbons. These U.S. laws address waste disposal and pollution prevention, control, and cleanup across geographic boundaries (by air, water, and soil) by setting science-based criteria and technology-based limits at the federal level, and use command and control or more flexible compliance methods (e.g., cap and trade incentives). Various levels of delegations are shared with state and local authorities. In addition, states may have delegated or parallel requirements.

In 1976, in the wake of a national hazardous waste crisis, Congress fundamentally changed the way solid and hazardous waste is managed in the United States by enacting RCRA.[4] RCRA, implemented by U.S. EPA and the states, created a "cradle to grave" solid and hazardous waste management system. This hazardous waste management system prohibited the previous practice of open dumping and replaced it with requirements to use engineered and regulated landfills, composting, and

---

[4]Resource Conservation and Recovery Act (RCRA), Public Law 94-580, October 21, 1976 (42 U.S.C. §§ 6901-6992; 90 Stat. 2795), as amended by P.L. 95-609 (92 Stat. 3081), P.L. 96-463 (94 Stat. 2055), P.L. 96-482 (94 Stat. 2334), P.L. 98-616 (98 Stat. 3224), P.L. 99-339 (100 Stat. 654), P.L. 99-499 (100 Stat. 1696), P.L. 100-556 (102 Stat. 2779).

recovery systems such as recycling.[5] RCRA has management requirements assigned to either "solid waste" or "hazardous waste" and currently treats plastic waste as a subset of "municipal solid waste" for disposal in landfills or by incineration.

Other U.S. environmental laws focus on preventing, controlling, and cleaning up discharges of pollutants, hazardous substances, and other contaminants to air and waters (including coastal and marine waters). These include laws enacted to control the discharge of pollutants or hazardous substances from certain facilities into the environment, such as the Clean Water Act, Clean Air Act, Ocean Dumping Act, and the Toxic Substances Control Act. In 1980, Congress assigned liability for cleanup and compensation for injury and contamination from historic contamination by enacting the Comprehensive Environmental Response, Compensation, and Liability Act (CERCLA, also known as Superfund). All of these laws are implemented by U.S. EPA as the lead agency. U.S. Coast Guard and NOAA have major roles for cleanup, removal, and damage assessment for injury in coastal and marine environments.

Neither the Clean Water Act nor the Clean Air Act controls or measures releases of plastic waste from littering, mismanaged waste, sewage outfalls, runoff, industrial emissions, or other sources. The legal or regulatory definitions of "pollutants" or "hazardous substances" do not include plastics or plastic pollution, though legal challenges are testing whether some may be included based on toxicity or other regulatory criteria. No specific plastic effluent limits for industrial wastewater, stormwater, and plastic production facilities exist unless established under a Clean Water Act regional protocol to protect certain receiving waters from specific discharges, such as from stormwater systems. These include Total Maximum Daily Load (TMDL) limits for "trash" in local water bodies in various locations. While these TMDLs are not specific to plastics, plastic waste is included in trash. The state of California has set plastic discharge limits to govern pre-production plastic discharges.

NOAA plays a leading federal role in plastic waste prevention, removal, cleanup, and restoration through a range of environmental authorities including the Clean Water Act and Ocean Dumping Act, which relates to ship-based disposal. Its most comprehensive role on ocean plastic waste is under the 2006 Marine Debris Research, Prevention, and Reduction Act, amended in 2012, 2018, and 2020 (Marine Debris Act), which specifies its role in cleanup, government coordination, grantmaking, and research. The Marine Debris Act does not provide specific authority for any federal agency to regulate the production, transportation, or release of plastic waste. The most specific legislative action around plastic

---

[5]Code of Federal Regulations (CFR) Title 40, Parts 239–282.

pollution in aquatic and marine environments was the 2015 Microbead Free Waters Act, which prohibits the manufacturing, packaging, and distribution of rinse-off cosmetics and other products, such as toothpaste, that contain plastic microbeads. U.S. EPA operates the non-regulatory Trash Free Waters program, which engages with states and communities on pilot prevention projects.

Most information available on U.S. plastic waste amounts, management, and leakage derives from solid waste data collected by U.S. EPA under RCRA, with other data from NOAA's Marine Debris Program, import or export data, and some state and local research, cleanup, or pilot projects.

CHAPTER SYNOPSIS

The potential for mismanaged waste starts at the generation of waste (discarded materials), although reused or donated materials are not categorized as waste. With the scale of U.S. waste generation, there is an opportunity to reduce the amount of waste produced, both for the environment as well as the economy, given that all waste management activities take effort, money, energy, and often transportation. As indicated in this chapter, there are multiple paths by which waste can enter into the environment. The next chapters describe how leaked plastic waste travels through the environment and the ocean.

PRIORITIZED KNOWLEDGE GAPS

As illustrated throughout this chapter, there are few data sources to understand sources, types, and relative scale of plastic waste generated and disposed or leaked to the environment beyond MSW in the United States. Specifically, there is a lack of plastic waste data on industrial wastes including pre-production plastics and fibers, nonpoint sources of waste such as runoff, point sources, wastewater treatment outflows, and sludge applications.

Furthermore, direct measurements of plastic waste and leakage, in different geographic regions of the United States and urban/rural environments, are necessary to improve and better constrain source estimates from existing crude (order-of-magnitude) model-based estimates, as illustrated in the U.S. EPA data.

FINDINGS, CONCLUSIONS, AND RECOMMENDATION

Finding 4: The United States is the largest generator of plastic solid waste, by mass and per capita. Plastic product end-of-life disposal can be improved by enhancing the capability of municipal solid waste systems to collect, sort, and treat specific materials and products, and by considering end-of-life disposal in plastic material and product design and manufacture.

Finding 5: Although recycling is technically possible for some plastics, little plastic waste is recycled in the United States. Barriers to recycling include the wide range of materials (plastic resins plus additives) in the waste stream; increasingly complex products (e.g., multi-layer, multi-material items); the expense of sorting contaminated, single-stream recycling collections; and the low cost of virgin plastics paired with market volatility for reprocessed materials.

Finding 6: Chemical recycling processes that strive toward material circularity, such as depolymerization to monomers, are in early research and development stages. Such processes remain unproven to handle the current plastic waste stream and existing high-production plastics.

Finding 7: Compostable plastics may replace some products currently made with unrecyclable materials. However, successful management of compostable plastics requires widely available managed composting facilities and consumer awareness on product disposal in dedicated compost collection, neither of which exists today.

Conclusion 2: Materials and products could be designed with a demonstrated end-of-life strategy that strives to retain resource value.

Conclusion 3: Effective and accessible solid waste management and infrastructure are fundamental for preventing plastic materials from leaking to the environment and becoming ocean plastic waste. Solid waste collection and management are particularly important for coastal and riparian areas where fugitive plastics have shorter and more direct paths to the ocean.

Conclusion 4: The United States has a need and opportunity to expand and evolve its historically decentralized municipal solid waste management systems, to improve management while ensuring that the system serves communities and regions equitably, efficiently, and economically.

Conclusion 5: Although recycling will likely always be a component of the strategy to manage plastic waste, today's recycling processes and infrastructure are grossly insufficient to manage the diversity, complexity, and quantity of plastic waste in the United States.

Recommendation 1: The United States should substantially reduce solid waste generation (absolute and per person) to reduce plastic waste in the environment and the environmental, economic, aesthetic, and health costs of managing waste and litter.

4

Physical Transport and Pathways to the Ocean

Plastic waste has a complex life cycle, moving from waste sources along a variety of long or short, direct or convoluted paths (Alimi et al. 2018, Bank and Hansson 2019, Eriksen et al. 2014, Hoellein and Rochman 2021). The ocean is Earth's ultimate sink, lying downstream of all activities. Almost any plastic waste on land has the potential to eventually reach the ocean or the Laurentian Great Lakes. Major paths of plastics to the ocean are summarized in Figure 4.1. These include urban, coastal, and inland stormwater outfalls; treated wastewater discharges; atmospheric deposition; direct deposits from boats and ships; beach and shoreline wastes; and transport from inland areas by rivers and streams (Dris, Gasperi, and Tassin 2018). This chapter reviews the many pathways that plastic waste can take from land to enter the ocean.

In the course of transport, plastic waste may encounter mechanisms that sort particles by density, size, and other characteristics that, in turn, affect their subsequent transport, physical and chemical characteristics, and ultimate fate in the environment. These processes affect the storage, availability, and impact of plastic waste at locations in shoreline, nearshore, and offshore environments. Processes that sort particles and influence their transport along various pathways are described in this chapter. Transformations that affect their size, number, shape, chemical composition, and biological and physical reactivity are discussed further in Chapter 5.

Figure 4.1 lists major pathways and mechanisms that move plastic waste to the ocean. The pathways are broadly categorized as waterborne,

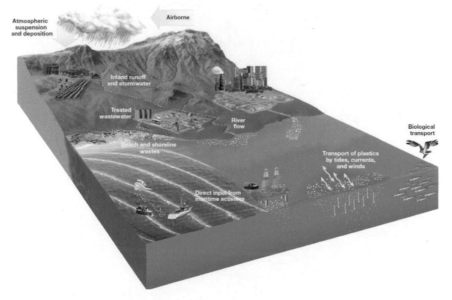

FIGURE 4.1 Major transport pathways for plastics from land to the ocean.

airborne, and direct deposit of plastic waste into the ocean. While the contributions of each pathway to the amount of plastic waste in the ocean are difficult to quantify, the following describes the state of knowledge about modes and patterns of transport, as well as examples of measurements or models of plastic waste transport along each pathway.

WATERBORNE PATHWAYS

Waterborne pathways of plastic waste include river flows, stormwater discharge, wastewater treatment plant effluent, and beach and shoreline wastes (see Figure 4.2). In the absence of a comprehensive U.S. national study, the presumptive pathway transporting the highest mass of plastic waste from both inland and coastal regions to the ocean is rivers and waterways. The mobilization of plastic waste along these pathways surges with floods and streamflow, as greater inundation gathers plastics from larger and more varied geographic areas and propels them seaward more energetically. These pathways also often bring about important transformations, delays, and barriers to the plastic waste they transport. Plastic particles' size, shape, and bulk density affect their waterborne transport (Haberstroh et al. 2021).

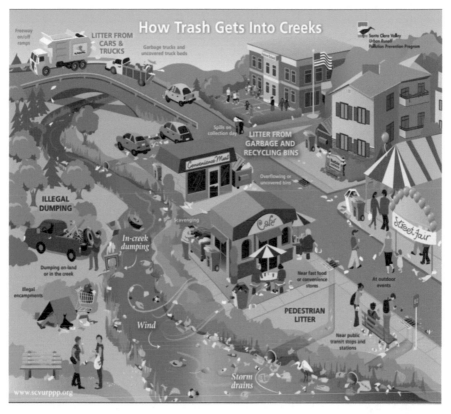

FIGURE 4.2 Sources and pathways of plastics in waterways. SOURCE: SCVURPPP (2021).

River Flow

Rivers and smaller waterways (e.g., streams, canals, channels) are major pathways for plastic waste entering the ocean from a variety of sources including littering (intentional or accidental), illegal dumping, and landfill leakage, as well as stormwater outfalls, combined sewer overflows, wastewater treatment plant effluent, and atmospheric deposition, which are described as pathways in more detail in subsequent sections (Williams and Simmons 1997, Windsor et al. 2019, Woodward et al. 2021). Once considered direct pipelines to the sea, rivers are dynamic drivers of plastic waste retention, burial, resuspension, and degradation as debris is transported downstream (Barrows et al. 2018, Hurley, Woodward, and Rothwell 2018, Nizzetto et al. 2016). Rivers, tributaries, and their floodplains are often "hotspots" of plastic accumulation (areas with the most marked and dramatic accumulation), with river and stream outlets also creating local hotspots in coastal marine areas (Windsor et al. 2019).

Plastic transport depends on the size, shape, and buoyancy of plastic items or particles, as well as river characteristics such as flow rate, velocity, and shoreline and waterway morphology (e.g., vegetation, rocks), which affect the time dependence of transport, including debris stranding and erosion (Balas et al. 2001, Hoellein and Rochman 2021). Variations in river discharge of plastics occur on a variety of time scales (Watkins et al. 2019), including those related to weather or climate variations (e.g., storm events, precipitation patterns) and source input (e.g., wastewater outflows or seasonal littering variability). For example, studies in the Los Angeles River (Moore, Lattin, and Zellers 2011) and the Chesapeake Bay (Yonkos et al. 2014) found debris concentrations increased sharply after major rainstorms. In Delaware Bay, local concentrations of floating plastics were driven by ocean tides and winds (Cohen et al. 2019), and in the River Seine (France), the mass of floating plastics increased with river flow (Gasperi et al. 2014, Tramoy et al. 2019).

Robust estimation of spatially and temporally variable transport (or flux) of plastic debris is rare. Across the globe, including in locations across the United States, the abundance of large plastic and microplastic debris in river water and sediments has been measured using a variety of methods (e.g., Adomat and Grischek 2021, Campanale et al. 2020, González-Fernández and Hanke 2017, González-Fernández et al. 2021). However, most studies report abundance at discrete sampling stations in one-time or short-term studies, potentially underestimating variability in time. For example, McCormick et al. (2016) measured the accumulation and export of anthropogenic litter from the riparian zone (up to 10 m from the water's edge) of rivers near Chicago, Illinois at biweekly and seasonal scales (McCormick et al. 2016). This riparian litter was highly mobile, a factor not captured in one-time "snapshot" sampling. Net accumulation rates depended on sampling frequency, where more frequent sampling gave higher accumulation rates. Also, they found that mobility varies with different debris characteristics. For example, because of their pliability, lightweight plastic films (wrappers and bags) were more likely to be retained on natural debris or vegetation than heavier, but more rigid, metal cans and glass bottles, which were transported farther. In some studies, microplastic loads increased after storm events (Yonkos et al. 2014) or periods of increased river discharge, and in one case the increase was attributed to combined sewer overflows (Wagner et al. 2019).

Quantitative global estimates of transport of plastic debris by rivers to the ocean come from modeling studies that use proxies including population density and mismanaged plastic waste generation rates to predict debris fluxes, which were then evaluated against available published data from a small number of individual rivers (Lebreton et al. 2017,

Schmidt, Krauth, and Wagner 2017). At least one field study found poor agreement between estimates based on field measurements and the previously modeled predicted outflow of plastics, in this case in six Chinese rivers (Mai et al. 2020). Meijer et al. (2021) added probabilistic modeling to account for the likelihood of land debris to enter a waterway as a function of distance from the shoreline, land use, wind, and precipitation. This study estimated that 0.8–2.7 million metric tons (MMT) of plastic waste enter the ocean globally per year from riverine sources, with 80% entering from more than 1,000 rivers. However, another study taking a similar modeling approach, but with slightly different model construction and calibration methods, estimated much smaller global plastic outflows from rivers (0.057–0.265 MMT [Mai et al. 2020]). There continue to be large uncertainties in the global estimation of riverine transport of plastic waste to the ocean, highlighting the importance of local field studies to more directly measure these fluxes and their variability. Such information will be valuable not only to better understand local sources and transport dynamics but also to build and validate models used for process studies and for regional or global budgeting studies.

Stormwater Runoff

Urban and suburban stormwater can be substantial and important contributors of plastic waste, especially microplastics from land to rivers and nearshore areas (Sutton et al. 2019). Stormwater runoff occurs when precipitation (e.g., rain and snowmelt) "flows over land or impervious surfaces, such as paved streets, parking lots, and building rooftops, and is not absorbed into the ground"(U.S. EPA 2021h). This runoff gathers debris and chemical pollutants, including plastic waste, from the land and streambanks (see Figure 4.3) and propels them to rivers, streams, lakes, and coastal waters, where they can harm humans and ecosystems (U.S. EPA 2020b).

Recent regulations on the amount of trash allowed in receiving water bodies in California have resulted in initial studies that estimate the total amount of trash, including plastic waste, generated and loaded in California's San Francisco Bay Area stormwater system (Werbowski et al. 2021). This study (discussed in greater detail later in the chapter) confirmed the findings of an earlier study in Los Angeles County, California, showing that trash loads could be roughly estimated by land use in the drainage area (EOA 2014). Researchers typically use land use as a proxy for stormwater trash loading in urban areas (Marais, Armitage, and Wise 2004).

The highest rates of plastic waste generation and loading found in California were from industrial, retail, and residential areas, as well as highways and expressways (EOA 2014). Other factors associated with higher plastic loading from urban areas include combinations of lower

FIGURE 4.3 Mixed stormwater debris. Photo credit: K. L. Law.

income, higher population density, and other demographic factors. However, significant correlations were not observed between generation rates and any individual factor. These results are similar to those for a study completed near Leipzig, Germany (Wagner et al. 2019).

Wastewater Discharges

Wastewater entering treatment plants can be highly contaminated with quantities of mostly fine microplastics, particularly fibers shed from clothing and other textiles (Carr, Liu, and Tesoro 2016). In wastewater treatment plants, most microplastics are removed and concentrated in wastewater sludges (Carr, Liu, and Tesoro 2016, Werbowski et al. 2021). These wastewater sludges are usually landfilled (buried), but some are applied to forest or agricultural land or are incinerated. Primary, secondary, and tertiary wastewater treatment removes cumulatively higher proportions and smaller sizes of plastic particles, with the remaining plastics discharged in liquid effluent, which may enter estuaries or the ocean directly, or rivers and streams leading to the ocean. Most microplastic removal occurs in primary treatment by surface skimming and sludge settling (Carr, Liu, and Tesoro 2016). Plastics in treated wastewater effluent tend to be much smaller in size and density and tend to be textile fibers or fiber fragments (Carr, Liu, and Tesoro 2016, Werbowski et al. 2021). Small amounts of these plastics can escape the wastewater collection system before they can arrive at wastewater treatment plants (e.g., during big storms which cause sewer overflows).

The role of onsite sewage disposal systems (e.g., septic tanks and cesspools) in the transport of microplastics to groundwater, and possibly to the ocean via groundwater discharge, is little studied. There is some preliminary evidence of the presence of microplastics in groundwater (Panno et al. 2019). Waterborne pollution delivery via submarine groundwater discharge to the ocean from onsite sewage disposal systems is known for other pollutants (e.g., Amato et al. 2020, and see Mezzacapo et al. 2021 for a state of Hawaii review). Coastal inundation events from storms, tides, or related to climate-induced sea level rise are expected to increase with attendant vulnerabilities to coastal- or waterway-adjacent onsite sewage disposal systems (Habel et al. 2017), potentially increasing the frequency of this type of microplastic transport. The importance of onsite sewage disposal systems as microplastic sources, and associated groundwater discharge of microplastics to waterways and the ocean, is presently uncertain.

Beach and Shoreline Waste

Beach and shoreline waste may be deposited as locally generated litter (accidental or intentional), or may be waste that was generated elsewhere and washed ashore (i.e., "beached"). The hydrodynamic processes that transport shoreline debris and determine its residence time before entering or returning to the ocean are complex. These processes are largely determined by local winds, waves, and tides, which are influenced by the shape of the coastline and seafloor bathymetry (van Sebille et al. 2015). The turbulence generated by wave breaking, especially in shallow areas such as the surf zone, can cause particles on the seabed or in sediments to be resuspended, and interaction of plastic waste with beach or seafloor sediments creates stresses that may enhance their fragmentation into smaller particles (Chubarenko et al. 2020, Efimova et al. 2018).

Delays and Barriers on the Way to the Ocean

Whether plastic waste entering inland streams is likely to arrive at the ocean depends on interceptions or transformations that occur along the way. This section examines processes that filter, sort, and delay plastics on the way to the ocean. Chapter 5 expands the discussion of chemical, physical, and biological transformations to plastic particles.

Sortings

Plastic particles transported by waterborne pathways often become sorted by density and size, much like natural sediments (Lenaker et al. 2019). Denser waterborne plastics tend to settle to the bottom, where they are transported as bedload sediment by river, storm, and tidal currents, and tend to deposit in bays, canyons, and nearshore areas (Barnes et al. 2009, Galgani, Souplet, and Cadiou 1996, Schwarz et al. 2019). Larger, denser particles tend to accumulate locally near river and stormwater outfalls, because stream velocities diminish in open water. However, very tiny (micron-sized) and more fibrous plastics tend to remain in suspension by fluid turbulence (Carr, Liu, and Tesoro 2016), causing them to move more readily in water flows (Liro et al. 2020, van Emmerik et al. 2018). For an individual plastic item or particle, this might lead to a cycle of transport, settling, and flood remobilization that prolongs its path to the sea for years (Liro et al. 2020).

In the nearshore region, highly periodic tidal currents are important in moving and sorting plastic particles. Plastic particles denser than seawater, such as tire particles, tend to settle but may continue to move under the influence of tidal and flood currents and may become resuspended by waves in shallower water (Chubarenko and Stepanova 2017, Sutton et al. 2019). Floating plastic particles, which are less dense than seawater, will tend to accumulate near the water surface and be moved by tidal and wind-driven currents. Particles near the density of seawater are expected to be suspended more evenly throughout the water column and be carried by ambient three-dimensional currents. The processes that affect the sorting, transport, and retention of plastic particles in coastal areas are complex and, thus far, little-studied (Sutton et al. 2019, van Sebille et al. 2015).

Filtration and Adsorption

Plastics can become stuck or filtered in ways that detain or retain them before reaching the ocean (see Figure 4.4). In particular, plastics are subject to contact with stream and river banks or floodplains, including vegetation, where they can become attached or deposited for a time or quasi-permanently (Ivar do Sul et al. 2014).

Biological Transport

Biological transport of inland plastic waste to the ocean or lakes, and vice versa, occurs via birds, fish, and other animals. Although the amount of transport by these means is likely small relative to the overall transport of ocean plastic waste, it can be meaningful from an ecological, community, or individual organism's perspective. The interaction of plastic waste and living organisms can result in negative impacts on organisms or ecosystems (Bucci, Tulio, and Rochman 2020). The nexus between biological transport of plastic waste and its distribution and fate is addressed in Chapter 5.

FIGURE 4.4 Mixed debris experiencing a delay during low flows on the Pearl River between Mississippi and Louisiana. SOURCE: U.S. Fish and Wildlife Service.

Microbial and other colonization of plastic waste in aquatic environments, also known as biofouling, can lead to the vertical transport of plastic waste in the water column (Tibbetts et al. 2018). Biofouling may alter the bulk density of plastic items, causing them to sink and affecting their settling.

AIRBORNE PATHWAYS: WIND

As with waterways, the atmosphere is both a transport mechanism and a reservoir for environmental plastics. Plastic waste from shed microplastics, to everyday litter (e.g., bags and wrappers), to large debris mobilized in severe windstorms can be suspended in the atmosphere and transported as a function of item size, density, and aerodynamic shape, as well as wind strength, turbulence, wind duration, and pathway obstructions. Figure 4.5 illustrates the familiar "Christmas tree effect" resulting from the snagging of plastic bags borne by wind in tree branches. Microplastics in soil, on roads, and at the ocean surface that are large enough to be entrained into the atmosphere and small enough to be elevated into the atmospheric planetary boundary layer can be subject to long-range transport and may have residence times up to 1 week (Brahney et al. 2021). Cycles of suspension, deposition, and resuspension (or emission and

FIGURE 4.5 Plastic bags caught in tree branches in Southport, Merseyside, UK. SOURCE: Shutterstock Stock Photo.

re-emission) of microplastics result in sizable reservoirs of atmospheric plastics. For example, the estimated average atmospheric load of microplastics (4–250 μm in size) over the land regions of the western United States is 0.001 MMT (Brahney et al. 2021). In analyses of particle pathways, research has demonstrated that microplastic transport can occur on regional scales (>100 km; see Allen et al. 2019) or can be dominated by large-scale (1,000 km) atmospheric patterns, resulting in deposition far from the emissions source (Brahney et al. 2021). Airborne pathways also carry a small proportion of microplastics resuspended from the ocean by sea spray for deposition on land (Allen et al. 2020, Brahney et al. 2021).

DIRECT INPUT

Plastic waste is also disposed of, either intentionally or unintentionally, directly into the ocean. These discharges include losses of fishing and aquaculture gear, recreational gear (e.g., during boating or scuba diving), over-board litter or intentional dumping, and cargo lost from ships and barges. Additionally, major storm events such as floods, hurricanes, and tsunamis can deposit massive amounts of debris of all types from land into the ocean in a relatively short time period. For example, the 2011 Tohoku earthquake and tsunami in Japan deposited an estimated 5 MMT of debris into the ocean (Murray, Maximenko, and Lippiatt 2018).

Finally, plastic particles that are shed during normal product use can directly enter the ocean. Examples include marine paints, coatings, and anti-fouling systems (International Maritime Organization 2019); shedding of textile fibers from synthetic clothing worn at sea; and shedding of particles from fishing gear (e.g., lines, nets).

CASE STUDY ON SAN FRANCISCO BAY AREA

San Francisco Bay is one of the most well-studied environments in the United States in regards to the transport and loading of plastic waste. Work discussed earlier on microplastic transport in the San Francisco Bay area provides some insight into quantification of flows into the ocean, in this case, from stormwater runoff and wastewater outflows (Sutton et al. 2019). The San Francisco Bay region also has a lengthy history of collecting trash data from beaches and inland shorelines during volunteer beach cleanups. An overview of the takeaways from these studies is provided below.

Despite being well investigated relative to other areas of the United States, the San Francisco Bay has important gaps in understanding of plastic waste transport and loading. Specifically, atmospheric deposition of microplastics has not been well studied.

Trash (All Types) Loading

A study focusing on the San Francisco Bay Area and Los Angeles examined debris that is captured in stormwater systems. The vast majority of collected debris was composed of organic material (e.g., vegetation), sand, and sediment. Trash (the debris composed of human-made materials) was 17% by volume and 4% by weight of all debris collected. Of the trash, plastics were roughly 70% by volume and 50% by weight (EOA 2014).

This study examined annual trash generation (all materials, not just plastics) and how loading to stormwater systems varies with land use, population density, and income (Figure 4.6; EOA 2014). As illustrated in Figure 4.6, trash loading rates vary up to three orders of magnitude between land-use classes, indicating that other factors must also be considered (EOA 2014). The reported units (gallons/year/acre) also illustrate the difficulty of standardizing units for reporting and analysis in this field.

The San Francisco Bay Region study was prompted by efforts to regulate trash in stormwater systems in California. These efforts are now being promulgated across the state due to recent amendments to state-wide stormwater permits that require municipalities and other entities to achieve zero discharge of trash into receiving water bodies. Recognizing

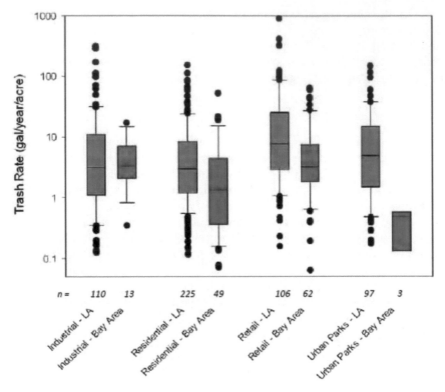

FIGURE 4.6 Ranges and median loading rates for all trash by land-use class for the San Francisco Bay Area and Los Angeles regions. The statistical minimum (lower whisker) and maximum (upper whisker), 25th percentile (lower box), median (horizontal line), and 75th percentile (upper box) are presented. Circles are statistical outliers as designated by the study. SOURCE: EOA (2014).

that achieving this goal will require effective monitoring methods by which to measure progress, the California State Water Resources Control Board and San Francisco Estuary Institute recently published the *California Trash Monitoring Playbook* in an attempt to help standardize data collection (S. Moore et al. 2021).

Microplastic Loading

Stormwater runoff of microplastics and microfibers is also an important contributor of plastic waste to coastal and near-coastal environments (Werbowski et al. 2021). While the volume or mass input may not be large due to the inherently small nature of the particles, the number of particles entering the marine ecosystem each year is extremely high. According to a recent study conducted by the San Francisco Estuary Institute and 5 Gyres, more than 7 trillion plastic microparticles and fibers enter the San Francisco Bay each year via stormwater runoff, which was approximately 300 times greater than the number of particles discharged by wastewater treatment facilities around San Francisco Bay (Sutton et al. 2019). Tire and road wear particles are a substantial component of synthetic microparticles to San Francisco Bay. This work can be used as a guide for interventions that target these sources.

Shoreline Debris from Community Science

In addition to the studies discussed above, the San Francisco Bay region has a decades-long record of community science efforts to capture data on beach and inland shoreline debris through volunteer beach cleanup efforts. Although the volunteer cleanups do not identify the ultimate paths of individual trash items, they indicate the types of items most frequently found in the environment, especially given the long-term data consistency. During the most recent year in which cleanups were held (e.g., not disrupted by COVID-19), 8 of the top 10 items found during cleanup activities were identified as plastics, comprising 67.3% of the total amount of debris collected (International Coastal Cleanup 2019). These beach cleanups identify common litter items (plastic and non-plastic) and can be used to inform litter prevention or mitigation efforts.

THE CHALLENGE OF ESTIMATING FLOWS OF PLASTICS ENTERING THE OCEAN

Although there is some understanding of the major mechanisms that transport plastic waste to the ocean, it is difficult to make quantitative estimates. Plastic waste inflows from each transport mechanism are very difficult to measure in the field. Inflows involve many large and small pathways and transport a very wide range of particle sizes, shapes, and densities, the smallest of which are often difficult to distinguish from natural fibers and materials. Furthermore, the fluxes vary over orders of magnitude with seasons, weather conditions, and location.

Another important challenge in assessing major paths and quantitative transport of plastic waste to the ocean is the lack of standard methods and data reporting within the scientific community. Each research team must decide what size debris to measure (micro–macro), the number of samples to collect and sampling area, the number of replicate samples to collect, and the time span between repeated sampling campaigns (if any) at the same site. In addition, studies report findings in variable units including mass (kilograms), particle counts (number of particles), and volume (gallons). For analysis of microplastics, specifically, one must select an extraction protocol to remove particles from tissues, organic matter, or sediment, as well as a method for chemical identification of some or all suspected plastic particles (Hidalgo-Ruz et al. 2012). Ideally, researchers make these decisions to best address their research objective, but cost, available resources, and other practical considerations are important. Researchers routinely call for standardization or harmonization of methods to ensure high-quality data and reproducibility between studies, and for reporting standards to allow robust comparability across local, regional, and global scales (Cowger et al. 2020, Rochman, Regan, and Thompson 2017). This is a priority for hypothesis-driven research and also for assessment and monitoring objectives. This lack of standardization in plastic waste studies has hindered the effective synthesis of current knowledge and is also discussed in Chapter 6 on Tracking and Monitoring.

KNOWLEDGE GAPS

A comprehensive understanding of the contribution of various transport pathways to plastic waste in the ocean is hindered by the complexity of the transport processes and thus the data needed to measure and model variability in fluxes over space and time. Improved understanding of the absolute and relative contributions of each pathway to plastic waste in the ocean could inform and prioritize actions to reduce the transport of plastics to the ocean. The committee identified the following research gaps needed to better understand transport of plastic waste to the ocean:

1. A lack of standardized or harmonized methods for measuring plastic and microplastic concentrations and fluxes hinders comparisons between data sets that are needed to make robust estimates at regional or global scales.
2. Without systematic field, laboratory, and modeling studies on processes influencing plastic and microplastic transport in water, in air, and on shorelines, flux estimates are necessarily crude, based upon limited field data that cannot fully capture variability associated with these complex processes. Such flux estimates are critical to both designing and implementing measures to reduce these fluxes, and to understanding the impacts of these fluxes. For example, identifying large mass inputs of plastic waste is important to inform the design of interventions to prevent transport into the ocean, whereas quantifying the abundance and potential toxicity of different microplastics transported to the ocean is critical to understanding the risk to marine organisms.

FINDINGS AND CONCLUSION

Finding 8: Although the transport of plastic waste to the ocean in the United States cannot be comprehensively estimated from available data, individual studies show a sizeable transport of microplastics and macroplastic wastes along a variety of waterborne and airborne paths as well as direct inputs from shorelines and maritime activities.

Finding 9: Plastic waste discharge to the ocean varies greatly with location and over time, reflecting variability in plastic waste generation by source, effectiveness of waste collection, and variability in transport processes such as river and stream flows; ocean waves, currents, and tides; and winds.

Conclusion 6: Regular, standardized, and systematic data collection is critical to understanding the extent and patterns of plastic waste inputs to the environment, including the ocean, and how they change over time.

5

Distribution and Fate of Plastic Waste in the Ocean

The distribution and fate of plastic waste in the ocean is a reflection of the amount and type of plastic waste that enters the environment from a diversity of sources, the efficiency of its transport from upstream locations to the ocean, and the transport and transformation of the material once it is in the ocean. For this report, the "distribution" of plastic waste is the concentration or abundance of plastics contained in a particular component of the ocean or the Laurentian Great Lakes, including coastal boundaries (Browne et al. 2015, Gray et al. 2018, Wessel et al. 2016), the water column (Choy et al. 2019, van Sebille et al. 2020, Woodall et al. 2015), the seafloor (Goldberg 1997, Williams, Simmons, and Fricker 1993), and within marine biota (e.g., Lusher et al. 2016). The "fate" is the final form of the plastic waste after undergoing physical and chemical transformations, and the permanent or semi-permanent location in the ocean dependent on this physical and chemical fate. Thus, the concepts overlap in defining the location of plastics within the ocean, though distribution may be a reflection of both short- and long-term storage occurring at any given time.

Transformation of plastic waste resulting from physical abrasion, photodegradation, chemical breakdown, or biodegradation will inform plastic waste life cycles, transport, and environmental sinks. This alteration of plastic waste is known to contribute to the generation of micro- and possibly nanoplastics as larger items are transformed ever smaller. The size of plastic waste greatly affects where it will be distributed in the ocean. Quantifying the rate of these transformations is a challenge described in this chapter.

In what form and where plastic waste resides determines its effects on natural, cultural, industrial, and recreational resources at local, regional, national, and global scales. Furthermore, understanding the distribution and fate of plastic waste is critical to informing mitigation strategies (described further in Chapter 7) such as cleanup and recovery options, understanding of global ocean plastic waste sources to achieve prevention, economic policies and other rulemaking, and citizen and consumer interest and engagement.

This chapter presents, synthesizes, and evaluates key information, where available, on the distribution and fate of plastic waste in the marine environment and Laurentian Great Lakes. It also identifies associated knowledge gaps and research opportunities, and reports associated findings. The chapter begins by examining estimates of plastic waste flows to the environment, which includes land, aquatic ecosystems, coastlines, and the ocean. It then describes the various reservoirs of plastic waste in coastlines and estuaries, the water column, seafloor, and aquatic life. Next, it explains the mechanisms involved in the transformation and ultimate fate of plastics in the marine environment. The final two sections present prioritized knowledge gaps and the committee's findings.

ESTIMATED PLASTIC WASTE INPUTS TO THE ENVIRONMENT

Table 5.1 summarizes estimates of plastic waste inputs to the environment, including land, aquatic ecosystems, coastlines, and the ocean, in the United States and globally. All estimates follow the basic modeling framework first presented in Jambeck et al. (2015), in which data on plastic waste generation and management are used to first estimate the amount of plastic municipal solid waste not collected in formal infrastructure (Jambeck et al. 2015). Nearly all studies, except for Lebreton and Andrady (2019) and Meijer et al. (2021), primarily used global municipal solid waste data compiled and reported by the World Bank. While Jambeck et al. (2015) estimated mismanaged plastic waste generated by coastal populations that entered the ocean, subsequent studies considered waste generated by populations living in inland watersheds, where mismanaged waste could enter and contaminate rivers and other waterways and ultimately reach the ocean. Studies focused on riverine input of plastic waste to the ocean included available (albeit limited) field data to calibrate and test their models (Lebreton et al. 2017, Meijer et al. 2021, Schmidt, Krauth, and Wagner 2017). Later models included additional pathways of plastic waste to the environment, including flows of microplastics (Lau et al. 2020) and export of plastic waste for reprocessing (Law et al. 2020), and Lau et al. (2020) also estimated the impact of the informal sector of waste collectors on the recovery of plastics with market value.

Estimates of global input of plastic waste to the environment vary by orders of magnitude, although few are directly comparable because of differences in modeling approaches, and none are grounded in extensive empirical measurements of plastic waste abundance or transport into the environment. However, these estimates do convey the scale of the problem, with up to 100 million metric tons (MMT) of plastic waste generated in a single year estimated to be uncollected in formal waste management systems globally. In the United States, despite a well-developed formal waste management system, approximately 1 to 2 MMT of plastic waste generated domestically was estimated to enter the environment at home and abroad (after export for recycling) in 2016 (Law et al. 2020).

ENVIRONMENTAL RESERVOIRS OF AQUATIC PLASTIC WASTE

There is an incomplete understanding of the distribution of plastic waste in aquatic (freshwater and seawater) environments, though the question is much investigated. For example, a recent scholarly review of the transport and associated distribution of floating ocean plastic waste cites 400 reference sources or studies (van Sebille et al. 2020). Since the ocean is a large and complex environment, it can be helpful to break it down into smaller components to better study and address plastic pollution at various spatial and temporal scales. These smaller scales can be considered reservoirs because they are regions where plastics are being held. Reservoirs considered in this report include coastlines and estuaries, ocean water column, seafloor, and marine life (Figure 5.1). This conceptualization necessarily involves some imprecision, for example, at the water column–seafloor interface and across stratified but contiguous water column depths. Furthermore, a comprehensive assessment of the amount of plastic waste in any particular environmental reservoir has yet to be achieved.

This section reviews a selection of the scholarly literature to illustrate and explore some of these reservoirs. Information and criteria related to each reservoir reflect its unique nature, as well as available data. This section does not present a comprehensive review of the literature, which continues to grow at a staggering rate. Chapter 4 describes inland reservoirs of plastic waste, which may remain in those areas and are thus not treated in this chapter.

The varying methods and units used across these studies make it difficult to understand the distribution of plastic waste in the ocean. The abundance of plastic waste is typically reported either as mass (weight) of items or as item count. Both measures are important and useful to inform strategies on ocean plastic waste. Mass budgeting is a tool used to assess stocks and flows of waste and is a sensible metric to assess the outcome of

TABLE 5.1 Estimates of Plastic Waste Inputs to the Environment, Including Land, Aquatic Ecosystems, Coastlines, and the Ocean, in the United States and Globally

Study	Estimate of Plastics Entering Environment (land, aquatic ecosystems, coastline, ocean)	Receiving Environment	USA	Global	Year of Estimate	MSW Not Collected in Formal Infrastructure	Illegal Dumping (USA only)	Littering
Jambeck et al. 2015	4.8–12.7 MMT	Ocean		☑	2010	☑		☑
	31.9 MMT	Coastline (50-km buffer)		☑	2010	☑		☑
	0.04–0.11 MMT	Ocean	☑		2010	☑		☑
	0.28 MMT	Coastline (50-km buffer)	☑		2010	☑		☑
Lebreton et al. 2017	1.15–2.41 MMT	Ocean		☑	2010	☑		Unknown
Schmidt, Krauth, and Wagner 2017	0.47–2.75 MMT	Ocean		☑	2010	☑		☑
	76 MMT	Land		☑	2010	☑		☑
Lebreton and Andrady 2019	60–99 MMT	Land		☑	2015	☑		☑
	0.0029–0.29 MMT	Land	☑		2015	☑		☑
Borrelle et al. 2020	19–23 MMT	Aquatic ecosystems		☑	2016	☑		☑
	0.20–0.24 MMT	Aquatic ecosystems	☑		2016	☑		☑
Lau et al. 2020	9.0–14 MMT	Aquatic ecosystems		☑	2016	☑		
	13–25 MMT	Land		☑	2016	☑		
Law et al. 2020	1.13–2.24 MMT	Land	☑		2016	☑	☑	☑
	0.51–1.45 MMT	Coastline (50-km buffer)	☑		2016	☑	☑	☑
Meijer et al. 2021	0.80–2.7 MMT	Ocean		☑	2015	☑		☑
	67.5 MMT	Land		☑	2015	☑		☑
	0.0024 MMT	Ocean	☑		2015	☑		☑
	0.27 MMT	Land	☑		2015	☑		☑

NOTE: This table represents best available estimates, which were made using data, methods, and assumptions that vary by study or source. Gray highlighted lines indicate estimates for the United States. MMT = million metric tons, MSW = municipal solid waste.

Microplastics Input	Informal Sector	Export of Waste	Entire Population	Population in Inland Watersheds (via rivers)	Coastal Population (50 km buffer)	# Countries Included (global estimates only)	Primary Data Source for Plastic Waste (MSW) Estimation
					☑	192 countries	World Bank (Hoornweg and Bhada-Tata 2012)
					☑		
					☑		
					☑		
				☑		182 countries	World Bank (Hoornweg and Bhada-Tata 2012)
				☑		233 countries	World Bank (Hoornweg and Bhada-Tata 2012); also Jambeck et al. 2015
				☑			
				☑		160 countries	Waste Atlas 2016; Hoornweg and Bhada-Tata 2012; also Jambeck et al. 2015
				☑			
				☑		173 countries	World Bank (Kaza et al. 2018); also Jambeck et al. 2015, Lebreton and Andrady 2019
				☑			
☑	☑	☑		☑		Unknown number of countries	World Bank (Kaza et al. 2018)
☑	☑	☑		☑			
		☑	☑				World Bank (Kaza et al. 2018); also USA-specific data
		☑			☑		
				☑		160 countries	Lebreton and Andrady 2019
				☑			
				☑			
				☑			

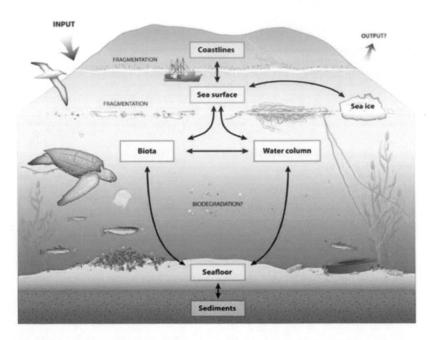

FIGURE 5.1 Schematic of plastic waste in the ocean and interactions that can occur from land to sea and from surface to seafloor. SOURCE: Law (2017).

source reduction activities. On the other hand, item count is more suitable for impact assessments, especially for microplastics, when the objective is to understand exposure to microplastics relative to natural prey during feeding, for example. Furthermore, abundance may be reported per unit area (e.g., mass or count per square meter or per square kilometer) or per unit volume (e.g., mass or count per liter or per cubic meter). In the absence of standardized field sampling protocols, each investigator appropriately determines the reporting unit(s) for their specific study. However, this creates difficulty when comparing results from different studies that followed different protocols and reported numerical data in different units.

The need for, and challenge of, defining standardized or harmonized (i.e., comparable) sampling and analysis protocols is commonly asserted in the scientific literature (e.g., GESAMP 2019, Hung et al. 2021), and researchers are working to evaluate existing methods (e.g., Hanvey et al. 2017, Löder and Gerdts 2015, Wang and Wang 2018) and to define guiding frameworks to collect data that would better inform risk assessments, for example (Connors, Dyer, and Belanger 2017). Until a time when such protocols may exist, researchers stress the importance of proper sampling design to address the stated scientific objective, strict quality assurance and quality control measures, and comprehensive reporting of methods utilized in studies quantifying plastic waste (especially microplastics) in the environment (Hermsen et al. 2018, Hung et al. 2021).

Throughout this chapter, the terms "abundance" and "amount" are used to describe quantitative measurements without specifying a particular unit. Interested readers should refer to original studies for further information about reported quantities.

Shorelines and Estuaries

Coastlines, including sandy beaches, rocky shorelines, and estuarine and wetland environments, are the recipients of plastic waste that may be generated locally, carried from inland sources (e.g., rivers, as described in Chapter 4), or brought ashore by storms, tides, or other nearshore processes. Microplastic and macroplastic waste, including litter and abandoned, lost, or otherwise discarded fishing gear (ALDFG), have been reported along coastlines worldwide, including in the United States. Historically, attention has been focused on litter found on sandy beaches (Browne et al. 2015), in part because of the decades-long International Coastal Cleanup (ICC) coordinated by Ocean Conservancy. Since the mid-1980s, when the first cleanup was carried out in Texas, citizen volunteers have participated in a 1-day annual beach cleanup on shores spanning the U.S. states and territories and more than 100 countries worldwide. In 2019, more than 32 million individual items were collected and categorized from more than 24,000 miles of beaches around the globe (International Coastal Cleanup 2020). The Top 10 list (highest number of items collected) has included the same familiar consumer products year after year, including cigarette filters, food wrappers, beverage bottles and cans, bags, bottle caps, and straws. In 2017, for the first time all items on the Top 10 list were composed of plastics (International Coastal Cleanup 2018). In 2013, in response to increasing attention to smaller debris, including microplastics, the category "Tiny Trash (less than 2.5 cm)" was added to the ICC data card.

The National Marine Debris Monitoring Program, which ran from 1996 through 2007 (and continued later under the National Oceanic and Atmospheric Administration's (NOAA's) Marine Debris Monitoring and Assessment Project, described in Chapter 6), was a federal beach monitoring program designed by the U.S. Environmental Protection Agency with support from other federal agencies and implemented by Ocean Conservancy, with goals to identify major sources of coastline debris and trends

in the amount of debris over time (Ribic et al. 2010). Regionally coordinated monthly surveys were conducted by trained volunteers to assess
the net accumulation of indicator items on beaches across the contiguous
United States, Alaska, Hawaii, Puerto Rico, and the U.S. Virgin Islands
(U.S. EPA 2002). An analysis of survey data (see Ribic et al. 2010, Ribic,
Sheavly, and Klavitter 2012) identified regional differences in amounts
and trends of land-based, ocean-based, and general-source debris that
were, in some cases, related to presumed drivers of debris sources including population size, land use, and fishing activity. The complexity of the
results of these scientific surveys is indicative of the challenges inherent
in assessing the amounts, sources, and trends of plastic waste in any
environmental reservoir.

More recently, Hardesty et al. (2017) reported an estimated 20 million
to 1.8 billion pieces of plastic debris along the shoreline of the United
States, based on a statistical analysis of beach data (average mass per mile
of shoreline) from the NOAA Marine Debris Monitoring and Assessment
Project, ICC data, and additional survey data collected for the project.
In this analysis several states were identified as national "hotspots" for
marine debris (see Figure 5.2), possibly related to coastal population density, urbanization (Mid-Atlantic states), transport by coastal currents and
wind patterns (Texas), and contributions from inland waterways.

The state of Hawaii is also particularly well known to suffer a disproportionately heavy marine debris burden, not only from locally based
marine litter (Carson et al. 2013) but also due to the state's mid-Pacific
Ocean location and associated exposure to widely circulated plastic pollution originating throughout the Pacific Rim (Donohue 2005, Ebbesmeyer
et al. 2012, Ingraham and Ebbesmeyer 2001, Kubota 1994, Matsumura and
Nasu 1997, McDermid and McMullen 2004, Moy et al. 2018). As a result
of oceanic convergence zones, aggregated debris of all types regularly
intersects the archipelago, including the northwestern Hawaiian Islands
that comprise the uninhabited and remote Papahānaumokuākea Marine
National Monument, an area of conservation and cultural importance
(Dameron et al. 2007, Donohue et al. 2001, McDermid and McMullen 2004,
Morishige et al. 2007, and see Howell et al. 2012).

The aggregation of plastic ALDFG in the nearshore waters and coastlines of the Hawaiian archipelago is particularly destructive as these
"ghost" gears and nets entangle marine life of commercial, cultural,
and environmental concern (Boland and Donohue 2003, Dameron et al.
2007, Donohue et al. 2001, Donohue and Foley 2007, Henderson 2001).
Fishing gear becomes abandoned, lost, or otherwise discarded for many
reasons such as adverse weather; gear conflicts; "operational fishing factors including the cost of gear retrieval; illegal, unreported, and unregulated fishing; vandalism/theft; and access to and cost and availability of

shoreside collection facilities" that may incentivize deliberate at-sea disposal (Macfayden, Huntington, and Cappell 2009). Worldwide, industrial trawl, purse-seine, and pelagic longline fisheries are estimated to lose a median of 48.4 kt (95% confidence interval: 28.4 to 99.5 kt) of gear during normal fishing operations annually (Kuczenski et al. 2022). This estimate, based on fishing activity in 2018, did not include abandoned or discarded gear; other gear known to become derelict such as pots and traps, poles and lines, and driftnets/gillnets; or gear from nearshore and small-scale fisheries (Kuczenski et al. 2022). By percentage, a separate study estimated annual ALDFG worldwide at 5.7% of all fishing nets, 8.6% of all traps, and 29% of all lines (Richardson, Hardesty, and Wilcox 2019).

At least 46% of the debris (by mass) in the Great Pacific Garbage Patch, an area of ocean plastic accumulation in waters between California and Hawaii, is estimated to be ghost gears and nets (Kuczenski et al. 2022, Lebreton et al. 2018). In 2007, it was estimated that 52 metric tons of ALDFG accumulate each year in the northwestern Hawaiian Islands alone (Dameron et al. 2007); more current estimates are unavailable. Furthermore, plastic debris is known to increase the susceptibility of reef-building corals to disease (Lamb et al. 2018) and was recognized at least as early as 2001 as a threat to Hawaiian coral reef ecosystems (Donohue et al. 2001).

Marine debris on Hawaii's coastlines is not limited to ALDFG. A 16-year study from 1990 to 2006 on one small atoll islet at French Frigate Shoals in the Papahānaumokuākea Marine National Monument documented more than 50,000 marine debris items with an annual deposition ranging from 1,116 to 5,195 items per year (Morishige et al. 2007). Morishige et al. (2007) reported that more than 70% of these items were composed of plastics. Smaller plastics, including microplastics, are also increasingly known to be found on the coastline and nearshore Hawaiian Island environments with potentially dire effects (Gove et al. 2019, McDermid and McMullen 2004, Morishige et al. 2007). On Hawaii's most visited and populous island, Oahu, beach microplastic densities of up to 1,700 particles per square meter have been documented—among the highest worldwide on remote island beaches (Rey, Franklin, and Rey 2021).

Alaska coastlines are also a known reservoir for significant amounts of plastic debris (Merrell 1980, Polasek et al. 2017). As early as 1974, 349 kg of plastic litter per kilometer of beach was recorded on Amchitka Island in the Aleutian Island chain (Merrell 1980). In one study, 80 km of coastline in five national park service units in Alaska were cleaned of more than 10,000 kg of debris, the majority of which was composed of plastics (Polasek et al. 2017), a finding consistent with earlier seabed studies offshore of Kodiak Island, Alaska (Hess, Ribic, and Vining 1999). Plastic waste on Alaska beaches is often characterized by large, buoyant objects

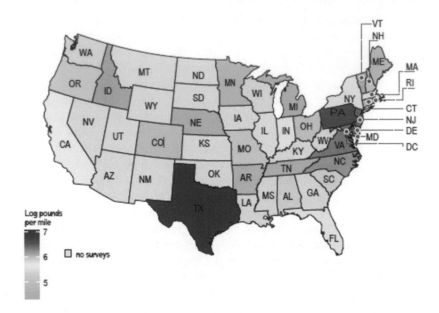

FIGURE 5.2 Assessment of debris load based on beach cleanup (International Coastal Cleanup) data for contiguous U.S. states (not including Alaska or Hawaii). Values represent the average weight of debris (including, but not limited to, plastics) per mile of shoreline for all coastal cleanup surveys across each state. SOURCE: Hardesty et al. (2017, Figure 2).

of maritime origin such as lines, buoys, and fishing nets that are likely wind- and current-driven to shore (Pallister 2012).

Several major estuaries and inland freshwater waterways in the United States have been surveyed for plastic debris, especially microplastics in the water column or buried in sediments (studies and their results detailed in Appendix D). These studies are widespread geographically—carried out in California (Los Angeles, San Francisco), the Pacific Northwest, and along the eastern seaboard from New York to Florida, as well as in regions far from the ocean (Illinois, Montana, Wyoming, Wisconsin, western Virginia). While these are a small number of studies relative to the number of estuaries and rivers in the United States, they have found microplastics to be ubiquitous.

In two estuary studies, particle concentrations were higher after rainfall or storm events (for San Francisco Bay, see Sutton et al. 2019; for Chesapeake Bay, see Yonkos et al. 2014), whereas in a study of the inland Gallatin River basin (Montana and Wyoming), microplastic concentration was inversely related to discharge, suggesting that microplastic sources are decoupled from discharge sources (Barrows et al. 2018). Studies in the metropolitan Chicago area found, at most sites, a higher microplastic abundance downstream of wastewater treatment plant outfalls than upstream, although no change in concentration was detected with increasing distance downstream (Hoellein et al. 2017, McCormick et al. 2016). A wide range of particle types, or forms, were reported across studies, with the proportion of fibers, fragments, pellets, films, and foams variable in each study. Some studies reported an abundance of polyethylene, polypropylene, and/or polystyrene, which are all polymers used extensively in packaging and other single-use applications. Direct comparison of results across studies is made difficult by differences in sampling methods, including particle sizes collected (dependent on net mesh size), sediment volumes or masses collected, and number of samples collected at a location (one-time versus repeated samples over time).

Plastic debris has also been well documented in the Laurentian Great Lakes, including in major Great Lakes tributaries, on shorelines, in surface water, and in benthic sediment (see systematic review by Earn, Bucci, and Rochman 2021). Individual studies reported plastic abundances comparable to or higher than those in ocean environments, with similarly large variation within studies due to environmental variability, and between studies due to different sampling and analysis methods (Earn, Bucci, and Rochman 2021). In one study, following similar methods to those estimating plastic input to the ocean (see Table 5.1), an estimated 10,000 metric tons of plastic debris from mismanaged solid waste entered the Great Lakes from the United States and Canada in 2010 (Hoffman and Hittinger 2017). In the same study, using a hydrodynamic model calibrated with field data, the authors identified likely accumulation zones across the Great Lakes and predicted the highest mass of floating plastic debris in Lake Erie (4.41 metric tons), followed by Lake Huron (1.44 metric tons) and Lake Superior (0.0211 metric tons).

Ocean Water Column

Floating Plastics

Some of the earliest reports of plastic debris in the ocean described small particles floating at the sea surface in estuarine (Kartar, Milne, and Sainsbury 1973), nearshore (Buchanan 1971, Carpenter and Smith 1972), and offshore waters of the North Atlantic Ocean (Carpenter et al. 1972) and large, identifiable objects (plastic bottles, balloon, sandal) floating in the open ocean of the North Pacific (Venrick et al. 1973). A more recent reanalysis of data from the North Atlantic Ocean and adjacent seas found that plastic contamination by large, entangling debris occurred as early as the 1950s, with significant increases observed in subsequent decades (Ostle et al. 2019). Since the 1970s, the majority of studies of the abundance and distribution of plastic marine debris have sampled the sea surface using plankton nets of varying types (van Sebille et al. 2015). The longest continuous data sets have been collected by undergraduate Sea Education Association Semester students sailing in the western North Atlantic since the mid-1980s (Law et al. 2010) and in the eastern North Pacific Ocean since 2001 (Law et al. 2014). The widespread coverage of surface plankton net data reported by a multitude of international research groups has allowed scientists to assess the large-scale accumulation of floating debris across ocean basins, which occurs in subtropical convergence zones centered around 30° latitude in ocean gyres in both the northern and southern hemispheres. These accumulation zones, commonly referred to as "garbage patches," are mainly composed of microplastics that have broken apart from larger items, although large floating debris (especially derelict fishing gear, including nets, floats, and buoys) is also found in these regions. The origin of these debris items (especially microplastics) typically cannot be determined except in rare instances, such as after the 2011 Tohoku earthquake and tsunami in Japan. Debris from this event, such as docks, vessels, and buoys, was identified for many years afterward, floating on the sea surface and washing ashore in Hawaii and North America (Carlton et al. 2017).

Contrary to common misperceptions of "garbage patches," floating plastic debris is not aggregated in a single large mass in the subtropical gyres but instead is dispersed across an area estimated to be millions of square kilometers in size (Lebreton et al. 2018). Even within the accumulation zones, particle concentrations (measured using plankton nets) can vary by orders of magnitude across spatial scales of tens of kilometers or less (Goldstein, Titmus, and Ford 2013), driven, at least in part, by physical transport processes creating small-scale convergences that are difficult to predict (see Figure 5.3).

Global estimates of the mass of floating plastics at the ocean surface have been made by synthesizing and extrapolating field data (Cózar et al. 2014), and with field data in combination with models of wind-driven ocean circulation to account for dispersal and variability across the ocean (Eriksen et al. 2014, van Sebille et al. 2015). Estimates vary depending on the data set used and data analysis methodologies, and range from 7,000–35,000 tons (6,350–31,751 metric tons) (Cózar et al. 2014) to 93,000–236,000 metric tons of microplastics (van Sebille et al. 2015), to 268,940 tons (243,980 metric tons) of microplastics and larger items (Eriksen et al. 2014) at the global ocean surface. All estimates of the mass of plastic waste in this sea surface "reservoir" have been only a small fraction of the estimated input of plastic waste to the ocean in a

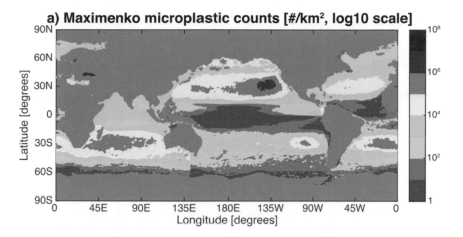

FIGURE 5.3 Map illustrates model prediction of microplastic abundance at the ocean surface. The highest abundances (warm colors) are in the ocean subtropical convergence zones (bands ~30 degrees latitude in the North and South Atlantic, North and South Pacific, and Indian Oceans), where ocean surface currents weaken and converge, causing floating material carried by the currents to accumulate. SOURCE: van Sebille et al. (2015).

single year (Jambeck et al. 2015). There are many possible explanations for this discrepancy. One explanation is the incomplete measurement of the size spectrum of floating plastic waste using plankton nets (which typically sample items from ~0.33 to 1 m) compared to visual observations by observers on ships or in aircraft, in which case only larger debris is detected because detection is dependent on the distance from observer to object. Furthermore, visual surveys are very resource-intensive and typically cover only small areas over short time periods. Bulk water samples filtered on very fine mesh filters have identified particles as small as 10 μm (Enders et al. 2015); however, sample volumes are very small and relatively few samples of this kind have been collected. Thus, the abundance and distribution of floating plastics across the known size spectrum (microns, and possibly nanometers, to many meters in size) is a major knowledge gap.

Suspended Plastics in the Water Column

Microplastics and occasional larger items, such as plastic bags, have also been detected in the water column between the sea surface and the seafloor. Vertical mixing of the water column driven by wind energy can distribute buoyant plastics to depths of tens of meters or greater (Kukulka et al. 2012, Reisser et al. 2015), and interactions with organic matter and biota may also cause initially buoyant particles to become dense enough to sink. A study in the nearshore environment of Santa Monica Bay, California (depths up to 15 m) found plastics larger than 0.333 mm at all depths sampled (Lattin et al. 2004), whereas a coastal survey off the U.S. West Coast only measured subsurface plastics (sampling to 212-m depth with plankton nets) in one out of four seasonal surveys (winter survey) (Doyle et al. 2011). Discrete water samples collected from remotely operated vehicles in Monterey Bay, California collected microplastics at 10 depths between 5 m and 1,000 m, with the highest concentrations (up to 15 particles per m^3) found between 200- and 600-m depth (Choy et al. 2019). In this study, the majority of microplastics were composed of polyethylene terephthalate (PET) and polyamide, polymers denser than seawater. Furthermore, eight discarded mucus feeding structures ("sinkers") of filter-feeding giant larvaceans and the gastrointestinal tracts of 24 pelagic red crabs examined in this study all contained microplastics. The known distribution and feeding behaviors of these animals are consistent with intake of microplastics between 100- and 200-m depth, indicating important interactions with organisms in pelagic ecosystems and a potential mechanism for vertical transport of microplastics to the seafloor (e.g., in sinkers).

Seafloor

Macroplastics and microplastics have been found in benthic environments around the world. Observed concentrations vary greatly, suggesting that source proximity and water currents and seafloor topography acting as concentrating mechanisms may play important roles in determining benthic loading.

Kuroda et al. (2020) conducted 63 surveys for seafloor marine debris in three areas of the waters off Japan between 2017 and 2019, using bottom trawls with 60- to 70-mm mesh nets at depths ranging from 67 to 830 m. The surveys identified debris concentrations averaging 2,962 items (53 kg) per km^2 in Hidaka Bay to 81 items (9 kg) in the East China Sea. Of all debris items, plastics accounted for 89% in Hidaka Bay, 69% off Joban, and 34% in the East China Sea. Based on information from labels on several debris items, Kuroda et al. (2020) estimated that about 30 years had elapsed between their manufacture and their retrieval from the seafloor. Comparison by Kuroda et al. (2020) to other studies in Japan and in Europe confirmed that plastics frequently account for the largest percentage of debris on the seafloor, though the percentage varies by location, from 22.2% (Hakata Bay, see Fujieda 2007) to up to 95% (eastern Mediterranean, see Ioakeimidis et al. 2014).

Peng et al. (2020) reviewed studies of the concentration of microplastics in seawater, beach sands and marine sediments, and marine biota. Abundance measurements varied greatly, which may provide a rough understanding of geographic variation, though they did not reflect standardized sampling methodologies, analyses, or units of measure. Smaller quantities have been detected in marine sediments in the Arctic (Kanhai et al. 2019) and Antarctic (Reed et al. 2018). Not surprisingly, higher, though variable, abundances are found near more populated areas (for Mediterranean Sea, see Guven, Gökdağ, and Kideys 2016; for North Sea, see Lorenz et al. 2019; for Plymouth, UK, see Thompson et al. 2004). Some studies that have measured microplastics in the water column and sediment have found higher abundances in sediments, suggesting the sediment is a significant sink for microplastics as they deposit over time (Kanhai et al. 2019, Song et al. 2019; however, see Zheng et al. 2019 for a reverse situation).

Nanoplastics have been identified in seawater using the presence of chemical markers (ter Halle et al. 2016), but their concentration and distribution have not been well resolved (Piccardo, Renzi, and Terlizzi 2020) as methods do not yet exist to directly detect and identify nanoplastics in the environment.

Benthic organisms may be impacted by exposure to deposited plastics and to toxic additives to the plastics. For example, hexabromocyclododecanes (HBCDs) are flame retardants commonly used as additives with expanded polystyrene and extruded polystyrene foam insulation, and with textile coatings. HBCD has been found in marine sediments (de la Torre et al. 2021, Klosterhaus et al. 2012, Sutton et al. 2019) and Laurentian Great Lakes sediments (Yang et al. 2012).

Marine Life

The intersection of the distribution of aquatic plastic waste, as well as its abundance, and freshwater and marine wildlife habitat use necessarily informs how and to what extent organisms encounter and entrain this pollution. The nexus of marine life and the distribution and fate of aquatic plastic waste has been illustrated through two primary mechanisms: ingestion/egestion of and entanglement in plastic waste by living organisms (Gall and Thompson 2015, Gregory 2009, Kühn, Bravo Rebolledo, and van Franeker 2015, Kühn and van Franeker 2020, Laist 1997, Shomura and Yoshida 1985). Ingestion is the taking in or consuming of food or other substances into the mouth or body. Egestion is discharging or voiding undigested food or other material, such as through feces or vomiting. One review by Kühn and van Franeker (2020) found documented cases of entanglement or ingestion by marine biota in 914 species from 747 studies—701 species having experienced ingestion and 354 species having experienced entanglement. When ocean- or lake-borne plastic waste becomes bioavailable to and is ingested by living organisms, they may serve as de facto vectors. As vectors, they could potentially distribute plastics through complex ecological mechanisms, such as foraging strategies, diurnal or seasonal movements, or via trophic transfer. The distribution and fate of ocean plastic waste thus both affects and is affected by the marine lifescape in ways not fully understood.

Ingestion of Plastics

The ingestion of plastic waste by aquatic life has been documented for hundreds of species (e.g., Figure 5.4, Kühn, Bravo Rebolledo, and van Franeker 2015, Kühn and van Franeker 2020). It occurs at spatial scales ranging from the planktonic ingestion of microplastics and nanoplastics

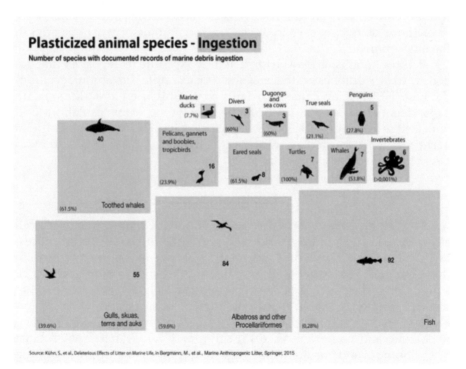

FIGURE 5.4 Visualization of the number of species with documented records of ingestion of plastics, based on a review of studies through December 2014. SOURCE: Maphoto/Riccardo Pravettoni, based on data from Kühn, Bravo Rebolledo, and van Franeker (2015).

(Botterell et al. 2019, Desforges, Galbraith, and Ross 2015, Lee et al. 2013, Sun et al. 2018) to the ingestion of all sizes of plastic debris by whales (Alzugaray et al. 2020, Baulch and Perry 2014, Besseling et al. 2015, de Stephanis et al. 2013, Im et al. 2020, Jacobsen, Massey, and Gulland 2010, Lusher et al. 2017, Unger et al. 2016). Nearly 60% of all whale and dolphin species have been shown to ingest debris with associated fatal results in up to 22% of stranded animals (Baulch and Perry 2014, de Stephanis et al. 2013, Jacobsen, Massey, and Gulland 2010). Figure 5.4 shows one global estimate of plastic ingestion by terrestrial, freshwater, and marine animals.

Entry of plastics into the ocean food web can occur when environmental plastics are consumed by organisms as a putative food source (Cadée 2002, Campani et al. 2013, Carr 1987, Lutz 1990, Mrosovsky, Ryan, and James 2009, Provencher et al. 2010, Ryan 1987, Schuyler et al. 2012, 2014, Tourinho, Ivar do Sul, and Fillmann 2010), via plastic-contaminated prey (Bourne and Imber 1982, Cole et al. 2016, Ryan 1987, Ryan and Fraser 1988), or indirectly through ambient seawater or sediments during foraging or other encounters (Beck and Barros 1991, Bravo Rebolledo et al. 2013, Di Beneditto and Ramos 2014, Murray and Cowie 2011). The interaction among such variables as availability of plastics in the environment, prey resemblance to plastics, prey selection, and the nutritional state of an organism has been hypothesized to increase the risk of plastic ingestion by individual organisms, a hypothesis largely supported by studies to date (reviewed by Santos, Machovsky-Capuska, and Andrades 2021). The preferential ingestion of plastics by some organisms has been shown to result from plastics' size, color, shape, age, abundance, or a combination of these factors (e.g., Botterell et al. 2019, Lavers et al. 2020, Lee et al. 2013). In certain seabirds, and perhaps other marine wildlife, plastic ingestion has been hypothesized to be facilitated by an olfactory signal—emanating from a complex biofilm that develops on aquatic plastic particles—that attracts birds to floating plastics (Savoca et al. 2016), though questions remain (Dell'Ariccia et al. 2017). When an organism's traits or behaviors become maladaptive in the face of environmental change it is termed an evolutionary trap; plastic ingestion has been identified as an evolutionary trap as a result of the availability of environmental plastics, plastics' mimicking of food options, and the proclivity of organisms to ingest plastics (Santos, Machovsky-Capuska, and Andrades 2021).

High concentrations of both microplastics and microscopic larval fish and invertebrates have been found in oceanographic features known as surface slicks, which are "lines of smooth water on the ocean surface" (Gove et al. 2019, Whitney et al. 2021). This discovery raises concerns regarding the trophic transfer of these plastics and associated toxins within the food web and ultimately to humans (Gove et al. 2019).

The presence of plastics in the excrement of secondary and tertiary marine predators has been offered as empirical evidence of trophic transfer in both captive (Nelms et al. 2018) and wild marine mammals (Donohue et al. 2019, Eriksson and Burton 2003, Le Guen et al. 2020, Perez-Venegas et al. 2018, and see Perez-Guevara, Kutralam-Muniasamy, and Shruti 2021 for a recent review of microplastics in fecal matter).

Plastic ingestion has been documented in the Laurentian Great Lakes, though a recent systematic review of the scientific literature demonstrates that the body of knowledge on plastics' effects on freshwater biota lags that which is known for the marine environment (Earn, Bucci, and Rochman 2021). Studies of the effects of plastics on freshwater biota have been predominantly laboratory based and hence not readily applied or extrapolated to the complexity of real-world conditions, among other caveats (Earn, Bucci, and Rochman 2021). Nonetheless, Earn, Bucci, and Rochman (2021) report that 60% of studies reviewed detected effects of plastics on freshwater biota (Earn, Bucci, and Rochman 2021). Notably, a recent study of fish in Lake Superior and Lake Ontario documented some of the highest abundances of microplastics and other anthropogenic particles in bony fish (marine or freshwater) reported to date (Munno et al. 2021). Of the two lakes, Lake Ontario fish had the greatest mean number of anthropogenic microparticles at 59 ± 104 standard deviations per fish and the greatest number to date in a single fish at 915 microparticles (Munno et al. 2021). Plastics in seafood being sold for human consumption have also been documented both in the United States (Rochman et al. 2015) and abroad (Naji, Nuri, and Vethaak 2018, Rochman et al. 2015, van Cauwenberghe and Janssen 2014), highlighting a potential route of trophic transfer of plastic contaminants to humans (Smith et al. 2018). Microplastics in particular have been identified as an emerging permanent contaminant of increasing concern in seafood (Farady 2019), though understanding of the relevance of this pollution to human health via seafood consumption is presently limited (Dawson et al. 2021).

Once entrained in aquatic food webs, within the bodies and tissues of living organisms across diverse taxa, plastic waste is subject to a diversity of spatiotemporal distribution mechanisms. An example is the transport of ingested plastic vertically in the water column through the diurnal vertical migration of zooplankton and fish, termed the "plastic pump." This plastic pump is also postulated as a mechanism by which plastics are delivered from shallower waters to the deep ocean including through fecal pellets (Choy and Drazen 2013, Cole et al. 2016, Katija et al. 2017, Lusher et al. 2016, van Sebille et al. 2020, Wright, Thompson, and Galloway 2013). As such, zooplankton have been postulated as a reservoir for microplastics (Sun et al. 2018), as have the water column and animals of the deep sea (Choy et al. 2019, Hamilton et al. 2021).

Animals that demonstrate high site fidelity to particular geographic locations, such as nesting or birthing sites, but ingest plastics during distant foraging may transport and distribute ingested plastics long distances upon their return (Buxton et al. 2013, Le Guen et al. 2020). The intergenerational transfer of plastics in seabirds that regurgitate ingested plastics to feed chicks has been known since the 1980s (Pettit, Grant, and Whittow 1981, Ryan 1988, Ryan and Fraser 1988). An additional example is the transport and distribution of microplastics by northern fur seals (*Callorhinus ursinus*) in the eastern North Pacific Ocean (Donohue et al. 2019). These seals forage offshore, returning to land in repeating cycles to rest, breed, or attend to their pups (Gentry and Kooyman 1986) and distribute microplastics ingested during foraging to novel locations, as feces containing microplastics are deposited on land (Donohue et al. 2019).

The biotic distribution of microplastics can also occur at smaller geographic scales, for example, through the sedimentary ingestion and subsequent concentrated egestion of microplastics by the sea cucumber (*Holothuria tubulosa*) (Bulleri et al. 2021). Bulleri et al. (2021) show that microplastic resuspension rates in the water column are greater from sea cucumber fecal material than surface sediments, facilitating microplastic bioavailability (Bulleri et al. 2021). While not an exhaustive treatment, the above examples demonstrate the diversity of taxa that may serve as reservoirs of ocean plastic waste and highlight the importance of considering marine life when addressing the distribution and fate of environmental plastic waste.

Entanglement in Plastics

The prevalence and distribution of ocean plastic waste is reflected in the ever-increasing number of species with plastic entanglement records—354 species by 2019, including birds, marine mammals, turtles, sea snakes, fish, and invertebrates (Kühn and van Franeker 2020, Kühn, Bravo Rebolledo, and van Franeker 2015; see also Figure 5.5). As with ingestion of plastics, studies of entanglement and other impacts of environmental plastics in freshwater systems have lagged those in marine systems, with assertions that freshwater impacts have been both underestimated and understudied (Blettler and Wantzen 2019). Entanglement in plastics, primarily derelict and operational/active fishing gear, has been identified as a primary threat to the endangered Hawaiian monk seal (*Neomonachus schauinslandi*) (Boland and Donohue 2003, Donohue et al. 2001, Henderson 2001) and North Atlantic right whale (Johnson et al. 2005, Knowlton and Kraus 2001, M. L. Moore et al. 2021, Myers and Moore 2020).

Entanglement of marine life in ocean plastic waste may distribute this pollution via the active or passive movement of living or dead entangled organisms across aquatic habitats, though the frequency and

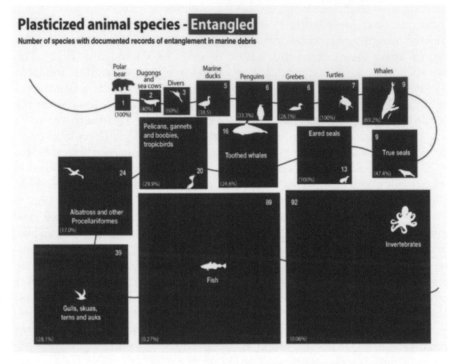

FIGURE 5.5 Visualization of the number of species with documented records of entanglement in plastics, based on a review of studies through December 2014. SOURCE: Maphoto/Riccardo Pravettoni, based on data from Kühn, Bravo Rebolledo, and van Franeker (2015).

ramifications of this mode of plastic waste distribution and transport are essentially unstudied. Scholarship has, understandably, focused primarily on understanding or documenting the effects of marine debris parameters (e.g., distribution, density) on individual species and biodiversity (e.g., Woods, Rødder, and Verones 2019). Seals entangled in derelict fishing gear are routinely observed returning to land with associated injuries such as deep and advanced wounds (Allen et al. 2012, Boren et al. 2006), suggesting they have been entangled for some time transporting the entangling net, line, rope, or other plastic waste with them. Individual North Atlantic right whales entangled in fishing gear are known to have carried the entangling debris on average at least 10 months and it is speculated that as the animals starve, lose body fat, and become denser, they sink at death, both concealing this marine debris-mediated mortality and distributing plastic debris to depth (Moore et al. 2006). In addition to the grave animal welfare issue entanglement presents (Butterworth, Clegg, and Bass 2012, Knowlton and Kraus 2001, Moore et al. 2006), the movement of ocean plastic waste by entangled organisms may also transport and distribute any living organisms present on the plastic waste, such as potentially invasive species (Kiessling, Gutow, and Thiel 2015, Miralles et al. 2018, Rech et al. 2018, Vegter et al. 2014) and novel viral or bacterial assemblages (Amaral-Zettler et al. 2016, Barnes 2002, Keswani et al. 2016, Kirstein et al. 2016, Masó et al. 2003, Zettler, Mincer, and Amaral-Zettler 2013).

Technical and design solutions to reduce entanglement threats to marine life have largely focused on reducing bycatch in actively fished (rather than derelict) gear (Hamilton and Baker 2019). Successful design advances include pingers (acoustic deterrent devices) for small cetaceans; exclusion devices for pinnipeds and turtles; and guard-type designs to reduce marine mammal entrapment in pots and traps (Hamilton and Baker 2019), though the effectiveness of these mitigation measures once gear become derelict is uncertain. Biodegradable panels on traps and pots have demonstrated success in reducing threats to marine life when traps and pots become derelict (Bilkovic et al. 2012). Some designs envisioned or proposed may ultimately reduce derelict fishing gear and associated entanglements, such as advances in gear marking (He and Suuronen 2018), ropeless trap and pot fishing (Myers et al. 2019), and biodegradable trap and pot panels (Bilkovic et al. 2012).

The Plastic Microbiome

Plastic litter can harbor unique microbial assemblages and may even facilitate the spread of antibiotic resistance across aquatic systems (Arias-Andres et al. 2018, Liu et al. 2021, Zettler, Mincer, and Amaral-Zettler 2013). Plastic microbial communities are distinct and more variable than those in the surrounding water and serve as effective disease vectors (Bryant et al. 2016, Kirstein et al. 2016, Lamb et al. 2018, Zettler, Mincer, and Amaral-Zettler 2013). Environmental DNA methods show the plastic microbiome may contain human and wildlife pathogens (Pham, Clark, and Li 2021). Most types of flotsam can serve as a vector of diseases and pollutants; however, the persistence of plastic litter and its transport and distribution exceed that of organic materials (Harrison et al. 2011). Multiple taxa associated with human gastrointestinal infections have been identified on microplastics downstream of wastewater treatments—but not in the surrounding water or organic matter—suggesting certain microbes may have an affinity for plastics (McCormick et al. 2016), though a recent comparative review of the science failed to confirm this (Oberbeckmann and Labrenz 2020). Members of the bacterial genus *Vibrio* are common in the plastic microbiome; many are harmless, but some are pathogens to humans and corals, and they frequently plague aquaculture facilities (Amaral-Zettler, Zettler, and Mincer 2020, Ben-Haim et al. 2003, Curren and Leong 2019, Lamb et al. 2018, Zettler, Mincer, and Amaral-Zettler 2013). Though understanding of the plastic microbiome remains incomplete, its role in microbial ecology underscores the diversity of impacts of ocean plastic waste.

Microbes may affect the distribution and fate of ocean plastic waste through colonization. Functioning as a microhabitat sometimes termed the "plastisphere," microbial colonization of aquatic plastic litter begins within hours and develops an amalgamated, crowded, complex three-dimensional structure of prokaryotes, archaea, protists, and detritus (Amaral-Zettler, Zettler, and Mincer 2020, Andrady 2011, Wright et al. 2020, Zhao et al. 2021). As this biofilm develops it can decrease the buoyancy of a microplastic particle forcing it to sink, thus enhancing its bioavailability (Andrady 2011, Eriksen et al. 2014, van Sebille et al. 2020). As mentioned previously, chemical signatures associated with biofilms on plastic waste may also serve as an attractant to foraging wildlife (Savoca et al. 2016). Microbial colonization, then, joins entanglement and ingestion-egestion as a biotic distribution mechanism for plastic aquatic waste.

TRANSFORMATION OF PLASTICS IN THE OCEAN

Two mechanisms are involved in the transformation and ultimate fate of plastics in the ocean: chemical and physical degradation, and potential for biodegradation.

Chemical and Physical Degradation

In the ocean, plastics are subject to wave and wind forces and solar radiation. Under these conditions, these plastics weaken and fragment into smaller and smaller particles (MacLeod et al. 2021). Physical degradation involves the breakage of bulk pieces of plastic into smaller fragments. Chemical degradation involves the breakage of chemical bonds in the plastic structure and may be accelerated by exposure to ultraviolet (UV) radiation, high temperatures, and elevated humidity (Chamas et al. 2020). This typically results in the creation of more microplastics and potentially nanoplastics that can accumulate in the ocean and be transported up the food chain through ingestion by fish, birds, and other aquatic species. Fragmentation into microplastics and nanoplastics increases the particle surface area, which facilitates the release of toxic additives into the environment (Arp et al. 2021). Despite the tendency to break into smaller pieces, plastics are known to have long half-lives, though specific degradation rates under various conditions are not well known (Chamas et al. 2020). The potential to degrade is dependent on both the plastic polymer type and the environmental conditions, which are most favorable at the ocean surface due to exposure to UV radiation, higher temperatures, and energetic waves.

Potential for Biodegradation

Nonmicrobial Marine Biota Transformation

As described in Chapter 2, both fossil-based and biobased plastics can have carbon-carbon bonds that require substantial energy to break apart. Degradation and, specifically, biodegradation depend on the chemical and physical structure of the plastics and the receiving environment, not where the carbon originates from. Therefore, biobased plastics are not necessarily more readily biodegradable than fossil-based plastics (Law and Narayan 2022).

While there are numerous records of ingestion of plastics by marine biota (described earlier in the chapter), there is a nascent understanding of the role marine biota may play in the transformation and ultimate fate of ocean plastic waste. In one study, microplastic particle size was not altered through the sedimentary bioturbation process (ingestion and egestion) of the sea cucumber (Bulleri et al. 2021). However, size reductions in ocean plastic debris have been observed in Antarctic krill (Dawson et al. 2018) and attributed to grinding of ingested plastics in the muscular gizzard of fulmarine petrel seabirds, followed by egestion (van Franeker and Law 2015).

Microbial Interaction with Plastics

Microbial utilization of plastics as a carbon (energy) source, possibly resulting in complete biodegradation (and removal) of the material, has been proposed. Recent work on ocean microbes has focused on characterizing the microbial communities found on ocean plastics compared to those on natural substrates and in free-living communities in seawater, and on understanding the interactions between colonizing marine microbes and specific polymers. As described in an earlier section, some of the first studies on marine microbes reported different microbial communities on plastics than on natural substrates or in seawater (e.g., Zettler, Mincer, and Amaral-Zettler 2013). However, in a recent critical review and comparative analysis of the scientific literature, Oberbeckmann and Labrenz (2020) found little evidence of polymer-specific microbial communities or of an increased affinity of pathogenic species for plastic substrates. Instead, they concluded that microbial communities on plastics tend to be opportunists that will readily colonize both synthetic and natural surfaces.

The vast majority of studies examining potential biodegradation of plastics in the marine environment (i.e., complete assimilation of plastic carbon by microbes and remineralization to CO_2, H_2O, and inorganic molecules) have focused on weathering (mainly photochemical degradation) and fragmentation (reduction in particle size) processes, which are necessary precursors to microbial assimilation and mineralization, particularly in the ocean (see Figure 5.6). However, relatively few studies have addressed microbial assimilation of carbon in traditional plastics to complete mineralization (removal) (Wang et al. 2018). Plastics with hydrolysable chemical backbones (e.g., PET and polyurethanes) may be more susceptible to enzymatic degradation and eventual biodegradation than those with carbon-carbon backbones (Amaral-Zettler, Zettler, and Mincer 2020), as illustrated by the discovery of PET-degrading bacteria

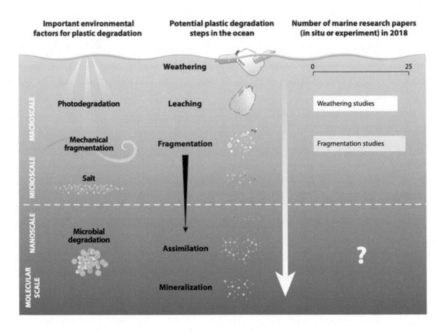

FIGURE 5.6 Schematic illustrating plastic degradation processes in the ocean and studied components. Vertical depth in this schematic indicates smaller sizes. The three columns across the schematic, from left to right, illustrate (1) factors for plastic degradation; (2) potential steps for degradation as particles become smaller; and (3) the available evidence to support each step across, from the year 2018. Steps underneath the white dashed line indicate processes that have not yet been validated in the marine environment. SOURCE: Oberbeckmann and Labrenz (2020).

isolated from a bottle recycling plant (Yoshida et al. 2016). However, Oberbeckmann and Labrenz (2020) argue, based upon Alexander's (1975) paradigm on microbial metabolism of a substrate, that the very low bioavailability and relatively low concentration of plastics in the ocean together with their chemical stability render these molecules very unlikely candidates for biodegradation by marine microbes, despite their potential as an energy and carbon source. Whether marine microbes have the evolutionary potential to adapt to plastic biodegradation in the future, especially if the concentration of plastics increases substantially in the ocean or in localized hotspots, remains an open question.

CHAPTER SYNOPSIS

A large and rapidly growing body of research documents the presence and characteristics of plastic waste throughout the marine environment, from the sea surface to seafloor sediments, coastlines to the open ocean, and in marine biota. The scale of plastic waste flows to the environment and the ocean has been estimated based on plastic waste generation rates and leakage outside of waste management systems, in the United States and globally. However, challenges remain in refining these global estimates and in identifying accumulation hotspots because of limited environmental data that are not readily comparable due to a lack of standardized methods, combined with large variability in ocean plastic concentrations in time and space. Addressing these knowledge gaps will improve estimates of plastic waste flows to the ocean from the United States and globally as a baseline from which to assess the impact of mitigation actions. Based on existing polymer chemistry and microbiology research, plastics (mainly carbon-carbon backbone polymers) are persistent in the marine environment, experiencing little to no biodegradation, and thus accumulate over timescales of decades or more.

KNOWLEDGE GAPS

There is insufficient information to create a robust (gross) mass budget for marine plastic waste and its distribution in ocean reservoirs. Measurements to date of plastic concentrations in individual locations over short time periods are difficult to extrapolate to larger areas and in time.

In order to improve understanding of the fate of plastics in the ocean, research is needed on the following issues:

1. The rate at which plastics physically and chemically degrade into smaller particles at various depths in the ocean, and how this varies by polymer type.

2. The fate of plastics in marine biota, including residence time, digestive degradation, and excretion rates.
3. The physical, chemical, and biological consequences of marine microbial interaction with different plastics.

FINDINGS AND CONCLUSION

Finding 10: Plastics are found as contaminants throughout the marine environment, including in marine life, but plastic amounts and volume in specific reservoirs or in the ocean as a whole cannot currently be accurately quantified from existing environmental data.

Finding 11: Research to date suggests that the distribution and concentrations of plastic waste in the ocean and Laurentian Great Lakes reservoirs can vary substantially across multiple spatial and temporal scales.

Finding 12: Plastics, especially those with carbon-carbon polymer backbones, are persistent and accumulating in the ocean. Even though plastics are chemically and physically transformed into smaller particles in the environment (e.g., through weathering-induced fragmentation and by interaction with biota), evidence suggests that biodegradation (complete carbon utilization by microbes) does not readily occur in the marine environment.

Conclusion 7: Without modifications to current practices in the United States and worldwide, plastics will continue to accumulate in the environment, particularly the ocean, with adverse consequences for ecosystems and society.

6

Tracking and Monitoring Systems for Ocean Plastic Waste

This report illustrates the limited, or absent, data from which to inform and implement effective plastic intervention actions. To inform source reduction strategies and policies, a national-scale tracking and monitoring program (or system of systems) is needed that spans the plastic life cycle—that is, from plastic production to leakage into the ocean (Figure 6.1). Tracking and monitoring plastic waste in the environment are essential to understanding and subsequently addressing the problem, but no comprehensive life-cycle tracking and monitoring of ocean plastic waste presently exists. Tracking and monitoring systems currently in place focus on solid waste management inputs and plastic waste items detected in the environment and ocean (Figure 6.1). This chapter explores tracking and monitoring systems currently in use and their limitations, and offers recommendations to inform the design, implementation, and benefits of a system or a system of systems to comprehensively track and monitor ocean plastic waste. Optimal systems will contribute to identifying and understanding the sources, transport pathways, distribution, and fate of ocean plastic waste, including legacy waste, to inform source reduction strategies or policies at multiple, if not all, intervention stages.

As noted in previous chapters, there are still immense gaps in understanding these processes, and there is an opportunity to utilize and expand tracking and monitoring programs to fill these gaps. Observational data are particularly valuable to inform scholarly modeling of plastic waste, such as mass-balance models that integrate and assess plastic material entering and leaving a system, as well as the fate of discarded plastics

FIGURE 6.1 Flow diagram of potential plastic waste interventions from plastic production to direct input into the ocean. SOURCE: Modified from Jambeck et al. (2018).

(Borrelle et al. 2020, Geyer, Jambeck, and Law 2017, Jambeck et al. 2015, Lau et al. 2020). Tracking and monitoring are two tightly related methods; in this report, tracking means following the transport of marine debris over time, whereas monitoring typically involves detection and measurement of plastic waste in the environment at various temporal and spatial scales. Most existing activities qualify as monitoring efforts. However, throughout the chapter, the committee refers to the value of both approaches.

Documentation of the extent and character of plastic waste and potential sources or hotspots (reservoirs and sinks) informs prevention, management, removal, and cleanup strategies (UNEP 2020). Moreover, it plays a critical role in evaluating the effectiveness of any interventions or mitigation actions, such as source reduction strategies or policies (described further in Chapter 7). Thus, information obtained through tracking and monitoring efforts is critical to share with the public and decision makers involved in motivating and designing intervention strategies.

There is no national-scale monitoring system, or "system of systems," to provide a baseline to track important sources, pathways, and sinks at the current scale of public or governmental concern. Under U.S. environmental management and protection law, monitoring systems are designed to achieve specific authorized purposes: legal compliance (e.g., waste generation or discharge monitoring), source detection (e.g., drinking water monitoring), and assessment of status and trends (e.g., ambient or in situ monitoring). The U.S. Environmental Protection Agency (U.S. EPA), the states, and other agencies operate a range of monitoring systems to meet such requirements, including those that monitor point and nonpoint sources and waste streams for pollutants or hazardous substances. These systems do not track or monitor plastic waste because it is not classified as a pollutant or constituent of concern. Much of the data on plastic waste is derived from data on municipal solid waste and, in a few cases, from nonpoint source trash monitoring, or from the efforts of research and community-based initiatives.

Part of the charge to the committee is to assess the value of a national marine debris tracking and monitoring system and how it could be designed and implemented. As specified in the task, this chapter considers how such a system may help in identifying priorities for source reduction and cleanup and assessing progress in reducing U.S. contributions to global plastic waste, and specifies existing systems and technologies that would be effective. The chapter gives particular attention to the National Oceanic and Atmospheric Administration's (NOAA's) Marine Debris Monitoring and Assessment Project (MDMAP), part of the NOAA Marine Debris Program (MDP), and potential improvements.

The chapter first explains existing tracking and monitoring strategies and programs. The following section describes considerations, enhancements, and opportunities for tracking and monitoring in the United States. The third section delves into the potential value of a national tracking and monitoring system. The final two sections outline priority knowledge gaps and present the committee's findings and recommendations.

EXISTING TRACKING AND MONITORING STRATEGIES AND PROGRAMS

Due to the lack of federal regulation of plastics as a pollutant in the United States and with the attendant lack of tracking and monitoring requirements, approaches to ocean plastic waste tracking and monitoring, including by the federal government, have been grounded in either research-based efforts or community science-based approaches.

Research-Based Approaches

Research-based monitoring for ocean plastic waste is often driven by government initiatives at various levels and geographic scales: local, regional, state, national, and tribal. One example is NOAA's MDP, which is directed by Congress to maintain an inventory of marine debris and its impacts. To help achieve this directive, NOAA's MDP offers several nationwide, competitive, short-term (<3 years) funding opportunities. Funds support "original, hypothesis-driven research projects focused on ecological risk assessment, exposure studies, and the fate and transport of marine debris" (NOAA Marine Debris Program 2021a). These projects may be conducted by government agencies, industry, or academic institutions.

Many local and regional research-based programs design their programs around concerns specific to that region. For example, plastic pollution is a central concern for the state of California. Among western U.S. regions, Southern California holds the greatest assemblage of plastic processors (Moore 2008), and California is the nation's most populous state with approximately 40 million citizens (U.S. Census Bureau 2019). As a microcosm of the national (and global) plastic pollution problem, California is leading research, removal, and prevention efforts.

While other U.S. states may not have the same focus and funding profile, lessons learned in California and other states can inform state and national efforts through research, removal, and prevention experiences.

Discrete, competitive ad hoc funding is appropriately employed to identify and fund hypothesis-driven research on aquatic plastic pollution but does not operate as a plastic waste tracking and monitoring system. While the information gained from this research can inform such a system, ad hoc research funding results in a disjointed monitoring record when individual projects end. This can contribute to a mosaic of plastic waste tracking and monitoring data collected using a diversity of methods, making it difficult to synthesize and interpret at meaningful spatial and temporal scales.

Community Science-Based Approaches

Community science-based approaches often include citizen-science activities or other experiential activities that also build public awareness and engagement. Experiential activities engage individuals through active participation, such as beach cleanups conducted through a variety of entities (often nonprofit organizations). Here, the term "community science-based" rather than "citizen science-based" is used to more accurately reflect the diversity of individuals engaging in the broader plastic waste tracking and monitoring enterprise. Community science-based efforts therefore may encompass citizen-science while recognizing diversity, seeking equity, and promoting inclusion.

A wide variety of community-based approaches are used to gather data on plastic pollution in the environment. Most approaches are focused on coastal areas, but a multitude of electronic mobile applications (apps) do not limit data gathering to coastal regions. This enhanced accessibility by a broader demographic has increased the transparency and availability of litter and other debris data along inland waterways and urban areas. The majority of these apps gather data and are not designed to answer specific research questions. The interpretation of those data to answer specific questions occurs *a posteriori*; therefore, the available data may not always be suitable to the questions. Furthermore, community science-based approaches do not routinely select locations in a scientifically rigorous manner, and thus the data collected may not be representative of plastic pollution at regional or national scales. Despite these limitations, several of these systems have been consistently gathering data on plastic waste for many years at various temporal and spatial scales.

A recent river basin-scale community science-based project illustrates the integration of community-based data collection with targeted research data collection in three pilot communities along the Mississippi

River (Youngblood, Finder, and Jambeck 2021). Researchers engaged the public in data collection using consistent transect-based methods so that the data could be compared with data from other research-based work in urban and riverine systems. The distinction between research- and community-based approaches is often blurred, and there is increasing interest in integrating research- and community-based science approaches (e.g., Earp and Liconti 2020, Liboiron et al. 2016). As with the Mississippi River project (Youngblood, Finder, and Jambeck 2021, NOAA Marine Debris Program 2021b), tracking and monitoring efforts may provide volunteers with specific research question-derived protocols that are distinct from cleanup-type protocols or opportunistic debris sightings used in other cases.

Selected Examples of Tracking and Monitoring Efforts

The following examples of plastic waste tracking and monitoring efforts are not intended to be comprehensive. Rather, they illustrate various approaches at assorted spatial and temporal resolutions. They may also potentially be integrated into a national-scale marine debris tracking and monitoring network or system of systems.

NOAA's Marine Debris Monitoring and Assessment Project

NOAA's MDP operates the MDMAP, a federal marine debris (plastics and other waste) inventory called for under the Marine Debris Act. MDMAP is the flagship community-science initiative of the MDP, engaging partner organizations and volunteers in a national shoreline monitoring program. The program has met many important national goals, including raising the issue to the public and decision makers, informing understanding of the risk and extent of marine debris in coastal and ocean areas, and identifying cleanup and mitigation priorities. Data collected and shared through the MDMAP are also intended to foster capacity at the local level in developing marine debris mitigation strategies to reduce impacts (NOAA Marine Debris Program. 2020b).

The foundation of MDMAP surveys is the NOAA-developed set of shoreline monitoring protocols (Lippiatt, Opfer, and Arthur 2013, Opfer, Arthur, and Lippiatt 2012) that standardize marine debris monitoring for consistent assessment of marine debris status and trends. The MDMAP surveys occur every 28 days (±3 days), as close to low tide as possible for shoreline sites that meet NOAA's criteria (i.e., sandy beach or pebble substrate, year-round access, no breakwaters or other structures that may affect coastal circulation, and no known regular cleanup activities). In 100-m-long sections, shoreline sites are surveyed for debris larger than

2.5 cm. Monitoring protocols include two shoreline survey types: standing stock and accumulation. Standing stock surveys are rapid visual assessments of debris concentration at a shoreline site. Accumulation surveys are tactile assessments that provide estimates of the flux, or accumulation rate, of debris at a shoreline site. For standing stock surveys, the 100-m-long sections are divided into twenty 5-m-long transects that extend from the back shoreline barrier to the water's edge. Surveyors identify and record debris items within four replicate, randomly selected transects. For accumulation surveys, debris is identified and removed from the entire 100-m site. To date, there are 9,055 surveys at 443 sites that span 21 U.S. states and territories and nine countries.

Studies, such as Uhrin et al. (2020), have demonstrated the utility of MDMAP data to estimate marine debris abundance and temporal trends, while also identifying associated limitations. The most extensive study on the benefits and challenges of existing marine debris monitoring programs, including MDMAP, is provided by Hardesty et al. (2017). The study was a collaborative project among Australia's Commonwealth and Industrial Research Organization, the Ocean Conservancy, and NOAA's MDP to better understand marine debris within the United States. Example survey issues identified include the following:

1. **Spatial sampling**. Most of the United States is not covered by existing data. Accumulation data are adequate for the West Coast, but standing stock is limited to concentrated efforts (Hardesty et al. 2017).
2. **Temporal sampling**. Accumulation rates are driven by regional/local biogeophysical forcing as well as debris type, such that the 28-day (±3 days) sampling window might be insufficient (Hardesty et al. 2017, Smith and Markic 2013, Uhrin et al. 2020).
3. **Site selection**. Environmental and anthropogenic factors impact debris counts (e.g., distance to the nearest town, freshwater outfall, nearest river) but were not strategized/prioritized when designing a long-term monitoring program (Uhrin et al. 2020).
4. **Substrate type selection criteria**. Shoreline debris monitoring methods are not analogous for rocky shores, and thus limited data exist for these environments (McWilliams, Liboiron, and Wiersma 2018, Thiel et al. 2013).
5. **Number of survey participants**. A linear relationship exists between debris counts and the number of participants, such that some surveys could be severely underestimated if the volunteer threshold is not met (Hardesty et al. 2017, Uhrin et al. 2020).
6. **Characteristics of survey participants**. The quality of the data collected by community scientists can be equivalent to that collected

by professional researchers, though variability may exist, for one example, younger primary school students detecting more debris than secondary students (van der Velde et al. 2017).

A key shortcoming of MDMAP identified by Hardesty et al. (2017) was the lack of a comprehensive national baseline for debris densities along the coast. This hinders the ability to monitor change in general, as well as change in association with the implementation of new policies and other interventions. In addition to a nationwide baseline survey, Hardesty et al. (2017) suggested regular surveys be conducted every 5 to 10 years at strategically selected sites in addition to continued citizen science efforts at self-selected sites. Aspects of these recommendations (i.e., one protocol, two approaches—community science and a national survey) appear in the NOAA MDP 2021–2025 Strategic Plan.

The International Coastal Cleanup

Developed and launched in 1986 by the nonprofit Center for Marine Conservation (now known as Ocean Conservancy), the International Coastal Cleanup (ICC) volunteer effort grew from a small local cleanup in Texas to an annual international effort, engaging with people in more than 100 countries. The Ocean Conservancy leveraged its partnerships with volunteer organizations and individuals worldwide to expand toward the Ocean Conservancy's Trash Free Seas initiative.[1] A pioneer in citizen science, the ICC was notable from inception insofar as it asked participants not only to collect coastal litter but also to document it using a standardized data card.

The ICC is the longest-running and most consistent community science data set, proving itself useful in both research and discussions around decision making. The ICC has been collecting largely the same data set since 1988, with comparable data available on local, regional, state, and nationwide perspectives. This data set has been used to track the effectiveness of regulations on plastic pollution. For example, the data set was used to evaluate the impacts of beverage deposit return schemes in the United States and Australia, finding that states that have a beverage container deposit result in 40% fewer containers littered (Schuyler et al. 2018). Data for the ICC are typically collected on a paper data card; however, an app (Clean Swell) is now available that mimics the paper data card, albeit with limited items. The full ICC data card is also integrated into the mobile app Marine Debris Tracker (described below), and the Ocean Conservancy

---

[1]See https://oceanconservancy.org/trash-free-seas/.

and University of Georgia are now coordinating on data collection and management. Both apps allow for more widespread collection of plastic pollution data through the engagement of a broader public demographic.

Marine Debris Tracker

Launched in 2011, the mobile app Marine Debris Tracker was the first litter or debris tracking app developed and has the longest history of electronic data collection (Jambeck and Johnsen 2015). In addition, it is one of few applications and programs to allow complete open access to all data ever collected. The Marine Debris Tracker was originally sponsored by a grant award from NOAA to the University of Georgia. NOAA subsequently sponsored research work with the app, with other partners contributing over time. The app has been used for various community science-based projects, as well as education and research initiatives (Ammendolia et al. 2021, Martin et al. 2019, National Geographic 2021, Thiel et al. 2017, Youngblood, Finder, and Jambeck 2021, Youngblood et al. In Review). In 2019, the app became sponsored by the global financial services company Morgan Stanley to professionalize it in partnership with the National Geographic Society, but the University of Georgia independently maintains science and data management for the app.

The Marine Debris Tracker database provides insights on managing, compiling, harmonizing, and visualizing plastic pollution data because it is a harmonized background database that allows for the creation of customized litter lists for individual organizations that vary in individual items cataloged. The app's harmonization allows for combined data compilation and statistics to be completed on the entire data set. To date, approximately 4 million items have been cataloged with the Marine Debris Tracker, with 2.33 million items originating in the United States. For example, the Mississippi River Plastic Pollution Initiative collected data on more than 75,000 debris items by both researchers and community members over 3 weeks in April 2021. Marine Debris Tracker was an early example of the successful application and acceptance of app use in community science, and remains the foremost and most comprehensive extant plastic pollution app.

Supporting Plastic Waste Mitigation with Monitoring Data

Data integration between electronically collected databases can provide a more complete picture of plastic waste and marine debris in the United States. While integrating these databases is not trivial, it is possible. The current three largest electronic research and community science-based data sets in the United States—the ICC, Marine Debris Tracker, and NOAA's MDMAP—are not well integrated.

Growing online and wireless connectivity nationwide and worldwide is making community science-based tracking and monitoring of plastic waste in the environment increasingly accessible. Many existing data collection efforts already allow the data to be visualized in map form.[2] Increased accessibility of plastic waste data through visualization tools has the potential to engage a larger, more diverse sector of society in community science-based activities—such as data crowdsourcing—toward awareness of and solutions to the ocean plastic waste problem.

Example Efforts

California established its Water Quality Monitoring Council's Trash Monitoring Workgroup "to support current practices and advances in trash monitoring" (California Trash Monitoring Methods Projects 2021). This Trash Monitoring Workgroup is "developing data analysis and visualization tools aimed at assessing the effectiveness of policies and practices for limiting the amounts of trash in the environment" (California Trash Monitoring Methods Projects 2021). One outcome was the 2021 publication of the *California Trash Monitoring Methods and Assessment Playbook*, which provides an overview of the methods in use to monitor trash in the environment (M. L. Moore et al. 2021).

Monitoring waste transport through watersheds (i.e., waste transported from the source via freshwater rivers and other waterways to the ocean) offers a more comprehensive understanding of plastic waste sources to guide targeted interventions. A recent research-based effort in Japan has quantified plastic emissions into the ocean using microplastic and macroplastic observations, correlations between microplastic concentrations in rivers and basins, and a water balance analysis (Nihei et al. 2020). This analysis estimated plastic input from Japanese land to the ocean as 210–4,776 tons per year. This work has also produced a plastic emissions map (Figure 6.2), which allows more efficient and effective deployment of plastic interventions throughout the country with a scale of 1-km grid cells. However, Nihei et al. (2020) did not include higher flow conditions or wastewater treatment plant outputs in the analysis.

The United Nations' Economic and Social Commission for Asia and the Pacific's Closing the Loop program[3] seeks to reduce plastic waste entering the ocean. This program has four main components: a plastic pollution calculator, a digital mapping tool informed by monitoring efforts, local action plans, and resource sharing. The International Solid Waste Association and the University of Leeds have worked with the

---

[2]See https://mdmap.orr.noaa.gov/.
[3]See https://www.unescap.org/projects/ctl.

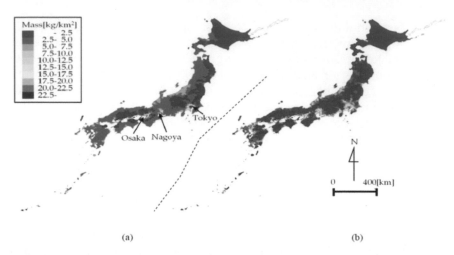

(a)   (b)

FIGURE 6.2 These maps indicate microplastic mass concentrations across 1-km grids in Japan for (a) population density and (b) urban area ratio. These estimated concentrations were found by use of linear approximation and a ratio of macroplastics/microplastics of 3.13. Hotter colors illustrate higher levels of microplastics, and cooler colors represent lower emissions. SOURCE: Nihei et al. (2020).

Closing the Loop program to use the Plastic Pollution Calculator to look at four cities to determine how plastics move from land to rivers and eventually to the ocean. The calculator provides information on sources, pathways, hotspots, and sinks of plastic waste to inform interventions to reduce ocean plastics. A digital mapping tool can examine images to determine the presence of plastic waste that could enter the ocean. This method can utilize images from a variety of sources, therefore reducing costs. The third component is creating a local action plan from the data gained from the plastic pollution calculator and the digital mapping tool. These plans are in process in Da Nang, Vietnam; Kuala Lumpur, Malaysia; Surabaya, Indonesia; and Nakhon Si Thammarat, Thailand. Last, a resource platform is being created along with an eLearning course to share information with stakeholders.

An additional program focused on Asia and the Pacific is Counter-MEASURE,[4] conducted by the United Nations' Environment Programme's Regional Office for Asia and the Pacific, which works alongside a variety of local and international partners. This work is funded by the Government of Japan. CounterMEASURE focuses on rivers as a source and transport mechanism of plastic pollution. CounterMEASURE has completed

---

[4]See https://countermeasure.asia/.

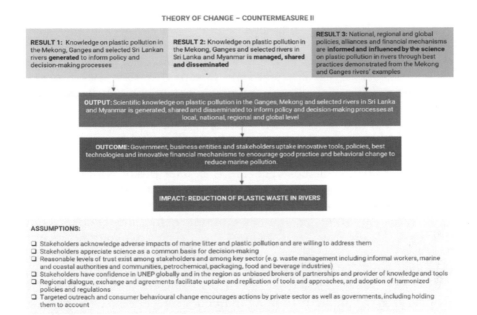

THEORY OF CHANGE – COUNTERMEASURE II

RESULT 1: Knowledge on plastic pollution in the Mekong, Ganges and selected Sri Lankan rivers **generated** to inform policy and decision-making processes

RESULT 2: Knowledge on plastic pollution in the Mekong, Ganges and selected rivers in Sri Lanka and Myanmar is **managed, shared and disseminated**

RESULT 3: National, regional and global policies, alliances and financial mechanisms **are informed and influenced by the science** on plastic pollution in rivers through best practices demonstrated from the Mekong and Ganges rivers' examples

OUTPUT: Scientific knowledge on plastic pollution in the Ganges, Mekong and selected rivers in Sri Lanka and Myanmar is generated, shared and disseminated to inform policy and decision-making processes at local, national, regional and global level

OUTCOME: Government, business entities and stakeholders uptake innovative tools, policies, best technologies and innovative financial mechanisms to encourage good practice and behavioral change to reduce marine pollution.

IMPACT: REDUCTION OF PLASTIC WASTE IN RIVERS

ASSUMPTIONS:

❑ Stakeholders acknowledge adverse impacts of marine litter and plastic pollution and are willing to address them
❑ Stakeholders appreciate science as a common basis for decision-making
❑ Reasonable levels of trust exist among stakeholders and among key sector (e.g. waste management including informal workers, marine and coastal authorities and communities, petrochemical, packaging, food and beverage industries)
❑ Stakeholders have confidence in UNEP globally and in the region as unbiased brokers of partnerships and provider of knowledge and tools
❑ Regional dialogue, exchange and agreements facilitate uptake and replication of tools and approaches, and adoption of harmonized policies and regulations
❑ Targeted outreach and consumer behavioural change encourages actions by private sector as well as governments, including holding them to account

FIGURE 6.3 Theory of change for the United Nations Environment Programme's CounterMEASURE Phase II program. SOURCE: CounterMEASURE (2021).

Phase I, which included the development of a conceptual framework for monitoring plastic pollution in rivers and a geographic information system data visualization platform, and is now expanding to Phase II to reduce plastic pollution in rivers regionally and globally. A description of CounterMEASURE Phase II, the "theory of change" to reduce plastic waste in rivers, is provided in Figure 6.3 and shows the interconnected nature of understanding the distribution of plastics and developing tools, policies, technologies, and innovative financial mechanisms to reduce marine plastic pollution.

CONSIDERATIONS, ENHANCEMENTS, AND OPPORTUNITIES FOR TRACKING AND MONITORING IN THE UNITED STATES

Spatial and Temporal Scales

The spatial and temporal scales of plastic waste data collection are very important because they will define the nature of the information gleaned from tracking and monitoring, as well as its potential usefulness in answering key questions. Data collected on marine debris items during coastal cleanups may illustrate waste management issues at local, regional, or national scales (Ribic, Johnson, and Cole 1997, Ribic, Sheavly, and Klavitter 2012, Ryan and Moloney 1993, Schuyler et al. 2018, Sheavly 2007, and see Ryan et al. 2009) but have been less effectively synthesized and interpreted at a global scale (Browne et al. 2015). Spatial monitoring of plastic waste is also commonly informed by elements of human geography such as the built environment, population density, and land use (Jambeck et al. 2015). Emerging technologies, described below, can expand our ability to collect data on plastic waste at a larger scale.

The timing of tracking and monitoring efforts will also shape the resulting findings. Widespread geographic monitoring at a "single" point in time can provide a static "snapshot" of aquatic plastic waste at various spatial or temporal scales; this type of monitoring is also known as standing stock sampling or standing stock surveys (Opfer, Arthur, and Lippiatt 2012, Ryan et al. 2009). Longitudinal sampling of locations at defined time intervals—ideally after initial cleanup—can provide dynamic information on plastic waste accumulation or reduction (Boland and Donohue 2003, Dameron et al. 2007, Morishige et al. 2007, Opfer, Arthur, and Lippiatt 2012, Ribic, Johnson, and Cole 1997, Ribic, Sheavly, and Klavitter 2012, Ryan et al. 2009), though sampling frequency may bias results from such factors as beach litter turnover or litter burial (Ryan et al. 2014).

A multitude of temporal factors may inform repeated sampling designs such as seasonality and the frequency and patterns of resource use such as beach attendance and fishing effort, among others (Jambeck et al. 2015). Opportunistic tracking and monitoring of ocean plastic waste associated with episodic or pulsed events such as tsunamis (e.g., Murray, Maximenko, and Lippiatt 2018), hurricanes/tropical cyclones (e.g., Lo et al. 2020), floods and precipitation events (e.g., Pasternak et al. 2021, Yu et al. 2002), or the capture of these events within established monitoring programs is also informative. When tracking and monitoring programs use standardized protocols, regional and site-specific comparisons are possible, greatly improving the ability of monitoring data to set priorities for source reduction and evaluate the success of intervention measures. To support site-specific and regional comparisons, NOAA developed standardized protocols and data collection for shoreline sampling (Opfer, Arthur, and Lippiatt 2012). Last, the scale of ocean plastic waste tracking and monitoring both in space and time is determined by the capital, including human capital, available and invested in such efforts.

Standardized Methods

Historically, data collection methods have been inconsistent among plastic waste tracking and monitoring efforts, resulting in detailed place-based studies but failing to form a body of research that can be compared geographically or temporally (Browne et al. 2015). Consistent methods used across geographic scales do allow for geographic comparisons, trend analyses, and data compilations. This has been possible through U.S. federal programs such as the National Marine Debris Monitoring Program, which ran from 1996 through 2007 (Ribic et al. 2010), and currently via the NOAA MDMAP (Hardesty et al. 2017).

Consistent, scientifically robust methods such as the use of randomized transects in cities, villages, and communities, often considered geographic sources of litter and leakage, are being used for projects to obtain data comparable across locations, over time, and in regional settings such as river basins (National Geographic 2021, Youngblood, Finder, and Jambeck 2021). In some cases, these methods are only applied by researchers. In other instances, community scientists with some level of training in the use of guiding tools such as mobile apps can meaningfully contribute to robust tracking and monitoring data collection. Participatory sensing of litter data can be opportunistic or led by research protocols. The latter improve data quality and facilitate the answering of specific research questions (Ammendolia et al. 2021, Jambeck and Johnsen 2015, Martin et al. 2019, Youngblood, Finder, and Jambeck 2021, Youngblood et al. In Review).

Development of standardized or harmonized (i.e., comparable) sampling and analysis protocols is a commonly asserted need, with known challenges (GESAMP 2019, Hartmann et al. 2019, Hung et al. 2021) that is gaining attention both in the United States and internationally. For example, an International Standards Organization (ISO) subcommittee on environmental aspects of plastics is currently working on standards to be used in a regulatory structure.[5] In the United States, U.S. EPA Region 9 is focusing on water quality monitoring methods and ASTM standards for sampling microplastics, which would enable microplastics to be included in the National Coastal Condition Reports and monitored in support of Clean Water Act § 303d impairment monitoring in states such as Hawaii and California; it could also be used in remediation and cleanup (Allen 2021). The state of California has already adopted a formal definition of microplastics for use in developing standards for drinking water and has developed a standardized methodology, sampling and analysis plan, health effects, and accreditation for drinking water by fall 2021 (California Water Boards 2021). Such standardization will allow for multiple tracking and monitoring efforts by researchers, communities, and industrial entities to be interpreted in aggregate.

---

[5]See ISO/CD 24187.2: Principles for the Analysis of Plastic and Microplastic Present in the Environment.

Study Design

The *a priori* definition of the purpose of a tracking or monitoring program is essential to effective program design. For example, monitoring for the quantity of plastics entering *the environment* differs from monitoring for the quantity of plastics entering *the ocean*. A first step in designing a monitoring system is often to articulate the questions to be answered through the establishment of the monitoring program. These questions guide the appropriate development and implementation of the monitoring program. In considering a design to address the entire life cycle of plastics (Figure 6.1), tracking and monitoring could occur from the production of resin polymers (the ultimate source of the material) through manufacturing, distribution, use, and disposal.

However, most often plastic monitoring is done at the waste management intervention stage and environment stages (Figure 6.1). This is often considered the *de facto* source of pollution because a majority of macroplastic pollution stems from mismanaged municipal waste. However, other pathways into the ocean exist, such as derelict fishing gear and direct input of microplastics from sources such as direct discharge, stormwater runoff, and tire wear, among others. Monitoring for leakage of waste can be used to pinpoint where the materials management system is disjointed or broken. Monitoring leakage of plastic waste could include measuring litter in cities, or along riverbanks or coastlines; capturing floating debris in rivers and waterways; or documenting plastics in the ocean. While leakage of plastic waste into the environment can be an indicator of a system that is not working properly, data further upstream in the plastic life cycle (e.g., production) can inform interventions that might have the most impact and be most cost-effective (Figure 6.1). In this role, tracking and monitoring can provide a more holistic understanding of the plastic materials management system toward enhanced and more informed policy-making and decision-making.

Some challenges related to designing a tracking and monitoring system include the following:

- inaccessible data, including proprietary data, which is why open, accessible data are so important;
- difficulty in collecting data over time for a large area such as the entire United States and its territories;
- limited data collection and analysis speed (which is improving with near-real-time data available from sites such as the Marine Debris Tracker);
- rapid and episodic changes in plastic use for which it is difficult to predict and plan monitoring (e.g., increased single-use plastic consumption and waste during the COVID-19 pandemic); and
- the ongoing degradation of larger plastic items or fragments into ever smaller pieces in the environment.

Given the degradation of plastics in the environment (see Chapters 4 and 5), there is a clear need for the identification, adaptation, or development of technologies to detect ever-smaller plastics. Current analytical practices are insufficient to detect environmental plastics at nanoscale sizes.

Available and Emerging Technologies

Intergovernmental agencies, environmental groups, and the research community have begun to assess all existing and emerging technologies for tracking and monitoring marine plastic debris, including *in situ* sensing, remote sensing, and numerical modeling, toward the goal of an integrated marine debris observing system (Maximenko et al. 2019 and depicted in Figure 6.4). These *in situ* sensing, remote sensing, and modeling initiatives could be integrated into already existing surface, inland, and coastal observing systems (e.g., NOAA's Integrated Ocean Observing System and state or federal water monitoring systems) and could form the basis for nationwide coordination around monitoring among different groups and using multiple technologies (similar to NOAA's National Mesonet Program for weather prediction). To do this effectively would require coordination between emerging technology programs and existing monitoring programs. Such coordination would focus on expanding collection measurements and protocols to allow remote sensing to measure plastic information already collected, GPS coordinates, photos, and, optimally, plastic spectra.

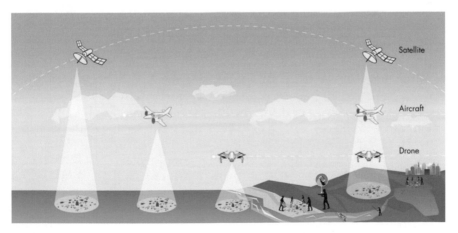

FIGURE 6.4 Depiction of a network of monitoring platforms that can be utilized as part of a marine debris observing system, collecting data at various scales.

Remote sensing has been emphasized as an underutilized and viable option for near-surface tracking and monitoring of plastic debris on land and at sea, and from land to sea (Figure 6.4) given the following: (1) the variety of available platforms (unmanned aerial vehicles or UAVs, aircraft, and satellites) and sensors; (2) its ability to provide spatially coherent coverage and consistent surveillance in time across scales—local to global (see also Martínez-Vicente et al. 2019); (3) its ability to access difficult-to-reach areas (Candela et al. 2021, Lavers and Bond 2017); and (4) its possibility to design a national monitoring program and illustrate where marine plastic debris is found (Candela et al. 2021).

Current remote sensing approaches under investigation with potential for marine debris detection include Synthetic Aperture Radar (Arii, Koiwa, and Aoki 2014, Matthews et al. 2017), bistatic radar (Evans and Ruf 2021), LIght Detection And Ranging (LIDAR) systems (Ge et al. 2016, Pichel et al. 2012), polarimeters, thermal infrared sensors (Garaba, Acuña-Ruz, and Mattar 2020, Goddijn-Murphy and Williamson 2019), and passive optical remote sensing (e.g., Acuña-Ruz et al. 2018, Biermann et al. 2020, Ciappa 2021, Garaba and Dierssen 2018, Goncalves et al. 2020, Kikaki et al. 2020, Topouzelis, Papakonstantinou, and Garaba 2019, Topouzelis et al. 2020). Assessment of the capabilities and limitations of remote sensing techniques are the subjects of active research (see Hu 2021, Martínez-Vicente et al. 2019, Maximenko et al. 2019). However, certain technologies have shown success in detection and thus could already be utilized as part of a tracking and monitoring system. Specifically, passive optical remote sensing is the most explored option with demonstrated potential in literature for inland, coastal, and open ocean marine debris detection (see Martínez-Vicente et al. 2019, Maximenko et al. 2019 for more information on all techniques).

Passive optical remote sensing includes red-green-blue (RGB) cameras, multispectral imagers, and hyperspectral imagers on various platforms (UAVs, aircraft, and satellites) with different spatial resolutions (on the order of submeter to hundreds of meters). RGB cameras simulate human eyesight, focusing on three bands within the visible portion (400–700 nm) of the spectrum. Multispectral imagers collect measurements in a limited number of wavelength bands (typically less than 10–15). Hyperspectral imagers (otherwise referred to as imaging spectroscopy) provide narrow, contiguous sampling across the spectrum (spectral sampling typically less than 10 nm translating to hundreds of wavelength bands). The spectral range covered by multispectral imagers and imaging spectroscopy is sensor dependent but can span the visible, near-infrared (NIR), and shortwave infrared (SWIR) spectral range (700–2500 nm).

RGB cameras on UAVs have been used extensively for indirect detection of marine litter on beaches and shorelines (e.g., Bao et al. 2018, Deidun et al. 2018, Fallati et al. 2019, Goncalves et al. 2020, Martin et al. 2018, Moy et al. 2018) with some application in coastal waters (e.g., Themistocleous et al. 2020, Topouzelis et al. 2020, Topouzelis, Papakonstantinou, and Garaba 2019), providing a cost-effective solution for localized image acquisitions at very high spatial resolution (on the order of centimeters). However, the practicality of RGB detection degrades as the platform changes to those at higher elevations, such as aircraft or satellite for regional to global coverage, wherein individual plastic targets will become less distinct with respect to their environment such that more wavelength bands are necessary to ensure accuracy between plastic debris and radiometric properties.

Recent laboratory studies revealed that marine plastic debris has unique spectral features in the NIR and SWIR spectrum (e.g., Garaba and Dierssen 2018, Hu et al. 2015, Knaeps et al. 2021, Moshtaghi et al. 2021, Tasseron et al. 2021). Therefore, passive methods that include the NIR and SWIR offer the greatest potential for direct plastic debris detection (Martínez-Vicente et al. 2019). Several recent papers have used NIR and SWIR spectral information from airborne imaging spectroscopy (Garaba and Dierssen 2018) and multispectral satellite imagery (e.g., Acuña-Ruz et al. 2018, Biermann et al. 2020, Ciappa 2021, Kikaki et al. 2020, Topouzelis, Papakonstantinou, and Garaba 2019, Topouzelis et al. 2020) to detect marine plastic debris in inland, coastal, and open ocean environments. Optical passive sensors provide an opportunity to identify and monitor leakage sources and accumulation regions (or hotspots), guide removal efforts, aid with the design or refinement of a national monitoring program (areas where field collection is a priority), and enable trend assessment over time with repeat observations.

Passive optical remote sensing has the potential to detect marine macroplastics at the ocean surface but likely not microplastics (from aircraft and satellite) and especially not at depth. For detection of microplastics, *in situ* methods have been applied to various environments, including marine and freshwater environments (e.g., Choy et al. 2019, Enders et al. 2015, Ghosal et al. 2018, Koelmans et al. 2019, Lenz et al. 2015, Tagg et al. 2015, van Cauwenberghe et al. 2013, Wolff et al. 2019, Zhang et al. 2017). Typically, water is sampled using bulk collection for small volumes or using plankton nets to filter large volumes, and samples are analyzed for potential plastic particles that must be identified via various techniques. Methods currently recommended by GESAMP (2019) for monitoring include optical identification (naked-eye detection, visual and fluorescence microscopy, and flow cytometry) and chemical identification/quantification methods (Fourier transform infrared [FTIR], Raman spectroscopy, pyrolysis-gas chromatography-mass spectrometry [py-GC-MS], and thermal extraction-desorption gas chromatography-mass spectrometry [TED-GC-MS]). See literature reviews from Araujo et al. (2018, Table 1), Mai et al. (2018), Primpke et al. (2020), Silva et al. (2018), and Zarfl (2019) for detailed information on all approaches and additional techniques (e.g., hyperspectral imaging, scanning electron microscopy), as well as sampling and sample extraction.

FTIR and Raman spectroscopic techniques (e.g., Araujo et al. 2018, Elert et al. 2017, Kappler et al. 2016) are the two most commonly used techniques to characterize microplastics and their polymers. The European Union expert group on marine litter recommended that all suspected microplastics in the 1–100 mm size range should have their polymer identity confirmed by spectroscopic analysis (Gago et al. 2016, MSFD Technical Subgroup on Marine Litter 2013). Within the literature, FTIR and Raman techniques have been used for analytical identification of microplastics ranging from biota, sediment, seawater, freshwater, and wastewater, to foods, beverages, and cosmetics (see Table 1 of Araujo et al. 2018 for a comprehensive list of Raman literature up to January 2018, and Primpke et al. 2020 for FTIR literature up to May 2019). The current limitation of Raman and FTIR imaging is the resource-intensive, both in time and dollars, nature of singular particle characterization.

Agency Coordination

Numerous agencies within the U.S. federal government have mandates or programs that directly or indirectly intersect with the issue of ocean plastic waste (U.S. GAO 2019). The value of interagency coordination has long been recognized, if not yet exhaustively achieved. The Marine Plastic Pollution Research and Control Act of 1987 (33 U.S.C. § 1914) (amending the Act to Prevent Pollution from Ships) provided for an "Interagency Committee," later amended by the Marine Debris Research, Prevention, and Reduction Act of 2006 (Marine Debris Act, 33 U.S.C. § 1954, as amended), to establish the Interagency Marine Debris Coordinating Committee (IMDCC). With the reauthorization and amendment of the Marine Debris Act by the 2020 Save Our Seas 2.0 Act (Public Law 115–265), the IMDCC remains a primary vehicle for enhanced interagency connectivity. Members include NOAA (which chairs the committee), U.S. EPA, U.S. Coast Guard, U.S. Navy, U.S. Department of State, U.S. Department of the Interior, U.S. Agency for International Development, Marine Mammal Commission, and the National Science Foundation.

The IMDCC serves as a legislated foundation for interagency coordination, including with regard to tracking and monitoring, but has unrealized potential in several areas, in part stemming from a lack of clarity on IMDCC membership (U.S. GAO 2019). The IMDCC has predominantly focused on its information-sharing role, citing the challenges of interagency collaboration such as mandate, mission, and budgetary appropriations variability among NOAA and other IMDCC members as barriers to expanded member coordination (U.S. GAO 2019). Research and technology development and coordination were among topics identified by experts in an audit report by the Government Accountability Office (GAO) of the IMDCC as areas of suggested action (U.S. GAO 2019). GAO suggested enhanced coordination among federal, local, state, and international governments and other nonfederal partners to address marine debris, as well as research on sources, pathways, and location of marine debris, inclusive of upstream elements such as rivers and stormwater. Tracking and monitoring environmental plastic waste is foundational to such efforts.

A national approach to tracking and monitoring mismanaged plastic waste that includes "upstream" source areas in the watershed has the potential to identify and inform intervention opportunities earlier, eliminating or reducing the time plastic waste is present in the environment. This necessitates enhanced collaboration and coordination with entities, including local, state, federal, and tribal agencies that have jurisdiction or other interests in the watersheds and waterways upstream of the coastal deposition of plastic waste. For example, the U.S. Geological Survey (USGS) maintains 27 regional Water Science Centers with core capabilities in hydrologic data collection, research and assessments, and information services. Their inland river and streamflow measurements, as well as flood forecasts, could inform aquatic plastic waste tracking and monitoring and potentially be co-located with plastic debris sensors as part of a monitoring network. USGS scientists have contributed to research-based monitoring and analysis efforts for microplastics (Baldwin, Corsi, and Mason 2016). A national approach may constitute a "system of systems," where programs and data collection efforts by various agencies, as well as research and community-based initiatives, are coordinated.

Effective Approaches to Tracking and Monitoring to Reduce Plastic Waste in the Ocean

Using their own experience and expertise, open session presentations from speakers, and research illustrated in this report, committee members created a list of tracking and monitoring program attributes expected to have the greatest efficacy in informing strategies to reduce plastic waste inputs to aquatic systems. Figure 6.5 illustrates a conceptualized approach to designing, implementing, evaluating, and adapting tracking and monitoring systems for plastic waste.

The following describes tracking and monitoring systems of plastic waste items expected to have the greatest efficacy in ultimately reducing

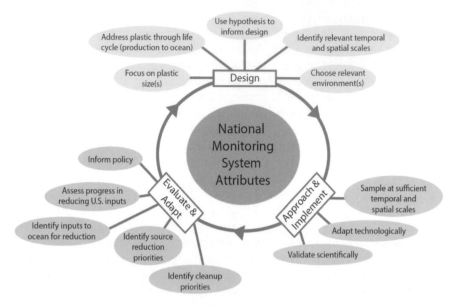

FIGURE 6.5 A conceptualization of the attributes of effective tracking and monitoring systems for marine plastic waste and other aquatic plastic waste. Even if all elements illustrated are not included, tracking and monitoring systems can still provide significant value based on specific needs, knowledge gaps, or other circumstances and are critical for the prioritization, design, and evaluation of interventions to reduce mismanaged plastic waste. Temporal and spatial scales are important to consider at the design stage and the approach and implementation stage. At the design stage, the focus may be on statistical power whereas the approach may have to include sampling changes in the field dependent on environmental conditions (e.g., weather).

plastic waste inputs to aquatic systems. The specific type or types of plastic waste addressed by any system, including polymer types, associated chemicals, or other characteristics or parameters of interest, will necessarily reflect the aims and drivers of those entities establishing the tracking and monitoring system.

- Tracking and monitoring systems that are **scientifically robust, hypothesis-driven, and conceptualized *a priori* to answer critical knowledge gaps,** rather than approaches applied *post hoc* to plastic waste tracking and monitoring questions.
- **Technologically adaptive tracking and monitoring systems** that can incorporate and utilize current and emerging technologies to

improve the spatial and temporal resolution of mismanaged plastic waste including the application of
- ○ remote sensing, autonomous underwater/remotely operated vehicles, sensor advances, passive samplers, and others;
- ○ crowdsourcing apps;
- ○ barcode tracking for recyclability and traceability;
- ○ biochemical markers and tracers that provide information on organismal exposure to environmental plastics, including legacy exposure and that which relates to organismal, including human, health; and
- ○ other current or emergent technologies.
- Tracking and monitoring systems that are **applied with sufficient spatial and temporal resolution** to capture meaningful data concerning knowledge and policy needs. For example, monitoring from a watershed perspective or including pre- and **post-intervention tracking and monitoring to assess progress.**
- Tracking and monitoring systems that **collect data that are comparable and, when scientifically robust, compatible with prior efforts. Examples include using standardized** measurement units or experimental design.
- Tracking and monitoring systems that **leverage, rather than separate, U.S. federal investment** in the reduction of mismanaged plastic waste among government departments and create synergies in the federal response to such waste.
- Tracking and monitoring systems that **encompass the full life cycle of plastics**, thereby achieving an understanding of the "upstream" plastic waste compartments and associated leakages.

POTENTIAL VALUE OF A NATIONAL MARINE DEBRIS TRACKING AND MONITORING SYSTEM

A single, national marine debris (or plastic waste) tracking and monitoring system does not exist in the United States, nor does such a system appear to be feasible given the complexity of plastic production, use, and disposal and the diversity of environments through which plastics are transported and distributed. A summary of marine debris/aquatic plastic waste tracking and monitoring systems and the intersection of such systems in addressing key aquatic plastic waste mitigation aims is provided in Table 6.1. This table illustrates that no single system or component serves as a comprehensive, stand-alone, national marine debris tracking and monitoring system. Furthermore, the specific aims of local, regional, national, and international efforts require the application of tracking and monitoring tools and technologies effective at particular spatial and temporal scales.

TABLE 6.1 A Summary of Marine Debris/Aquatic Plastic Waste Tracking and Monitoring Systems, Components, or Technologies and Their Intersection in Addressing Key Aquatic Plastic Waste Mitigation Aims

System, Component, or Technology	Size Class Sampled or Tracked	Identify Source Reduction Priorities	Identify Cleanup Priorities	Assess Progress in Reducing U.S. Inputs	Reduce Inputs to Ocean	Inform Policy
Community/citizen science/traditional and indigenous community cleanups	Micro Meso Macro					
Community/citizen science/traditional and indigenous community data collection and surveys	Meso Macro					
Industry/corporate efforts[a]	Micro Macro					
Municipal solid waste organizations and entities	Micro Macro					
Derelict fishing gear surveys	Macro					
Passive or static capture systems[b]	Macro					
Remote sensing applications	Macro					
Government/agency efforts[c]	Meso Macro					

Opportunistic systems or surveys of opportunity[d]	Macro				
Opportunistic and episodic events[e]	Micro Meso Macro				
Research-based systems[f]	Micro Meso Macro				

[a]For example, reporting of plastic production data and use by sector.

[b]For example, Mr. Trashwheel, retention booms, capture devices, stormwater structures, outflow pipe of wastewater treatment plant.

[c]For example, National Oceanic and Atmospheric Administration's Marine Debris Monitoring and Assessment Project, National Aeronautics and Space Administration, U.S. Environmental Protection Agency, the U.S. Geological Survey, government point and nonpoint source monitoring.

[d]For example, submersible missions, vessels of opportunities.

[e]For example, hurricanes/tropical cyclones, animal strandings, first-flush precipitation events.

[f]For example, institutes, colleges, think tanks.

NOTE: The degree of shading indicates the existing or potential value of the system, component, or technology in achieving a mitigation aim, with darker shading representing greater value. The size classes of plastic waste customarily addressed by each system, component, or technology are categorized as microplastics, mesoplastics, or macroplastics. Tracking and monitoring systems, components, or technologies are not presently available for environmental detection of nanoplastics (<100 nm in size) and are thus not included in this table. SOURCES: Koelmans, Besseling, and Shim (2015) and Mattsson et al. (2018).

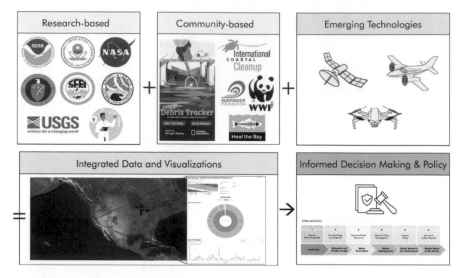

FIGURE 6.6 Depiction of the components of a national marine debris tracking and monitoring network, consisting of research and community-based initiatives, supplemented and supported by large-scale monitoring by remote sensing methods. Integrated data and associated visualizations would provide comprehensive understanding of plastic pollution in the United States, critical to informing actions toward plastic pollution reduction.

However, the use of multiple, complementary tracking and monitoring systems (depicted in Figure 6.6) in a synergistic approach implemented at sufficient spatial and temporal scales would contribute to (1) understanding the scale of the plastic waste problem and (2) the identification of priorities for source reduction, management, and cleanup and the assessment of progress in reducing U.S. contribution to global ocean plastic waste. For example, an optimal monitoring system design for first flush events would be useful to inform cleanup sites, track their progress, and reduce inputs to the ocean. The design could encompass community science cleanups, capture devices, trash booms, and remote sensing approaches.

KNOWLEDGE GAPS

Currently, data collected by various monitoring efforts are not well integrated. There would be significant value in developing a data and information portal by which existing and emerging marine debris/aquatic plastic waste data sets could be integrated to provide a more complete picture of the efforts currently tracking plastic pollution across the nation. Such a portal would need to be supported by (1) standardized methods of data collection and (2) support for long-term data infrastructure. The ability to visualize the data contained in the portal would greatly enhance its utility for the public and decision makers to inform and assess the progress of plastic waste reduction efforts.

FINDINGS AND RECOMMENDATIONS

Finding 13: No national-scale monitoring system, or "system of systems" exists to track important sources, pathways, and sinks of plastic waste to the ocean at the current scale of public or governmental concern. Presently, no baseline exists nor does a monitoring system to track changes from such a baseline.

Finding 14: The complexity of plastic production, use, and disposal, and the diversity of environments (inland to ocean) through which plastics are transported and distributed, requires the use of an expanded suite or network of tracking and monitoring systems to set priorities to reduce global ocean plastic waste.

Recommendation 2: The National Oceanic and Atmospheric Administration (NOAA) Marine Debris Monitoring and Assessment Project, led by the NOAA Marine Debris Program, should conduct a scientifically designed national marine debris shoreline survey every 5 years using standardized protocols adapted for relevant substrates. The survey should be designed by an ad hoc committee of experts convened by NOAA in consultation with the Interagency Marine Debris Coordinating Committee, including the identification of strategic shoreline monitoring sites.

Recommendation 3: Federal agencies with mandates over coastal and inland waters should establish new or enhance existing plastic pollution monitoring programs for environments within their programs and coordinate across agencies, using standard protocols. Features of a coordinated monitoring system include the following:

- Enhanced interagency coordination at the federal level (e.g., the Interagency Marine Debris Coordinating Committee and beyond) to include broader engagement of agencies with mandates that allow them to address environmental plastic waste from a watershed perspective—from inland to coastal and marine environments.

• Increased investment in emerging technologies, including remote sensing, for environmental plastic waste to improve spatial and temporal coverage at local to national scales. This will aid in identifying and monitoring leakage points and accumulation regions, which will guide removal and prevention efforts and enable assessments of trends.

7

Interventions for U.S. Contributions to Global Ocean Plastic Waste

The last component of the statement of task is to "recommend potential means to reduce United States contributions to global ocean plastic waste." In considering interventions for the United States, several themes emerge from expert advice on ocean plastic waste:

The need and ability to act without perfect knowledge. Government-led expert reports and scientific assessments from the United Nations (UN), the European Union (EU), Canada, the United Kingdom, Nordic countries, and U.S. states (e.g., California) advise precautionary and immediate action, from source reduction to reuse—even with existing uncertainties—while concurrently addressing key knowledge gaps (Brander et al. 2021, Environment and Climate Change Canada and Health Canada 2020, HM Government 2018).

The need for a systemic approach involving actions across multiple institutions. Expert reports from the UN (Cornago, Börkey, and Brown 2021, IRP 2021) and nongovernmental organizations (Ellen MacArthur Foundation 2017, Lau et al. 2020, World Economic Forum, Ellen MacArthur Foundation, and McKinsey & Company 2016) articulate the need for an integrated range of strategic interventions and advocate enforceable legal requirements and investments around waste prevention and management, product standards, and multisector commitments to reduce sources of plastic waste. Governments are aligning with "all of the above" principles and engaging in multisector collaborations for unified, systemic change.

The need for government and industry standards, goals, criteria, and rules to advance action. There is growing recognition that government goals, standards, and regulations are needed to enable coordinated action with industry and civil society to reduce plastic waste flows to the ocean. Although addressing plastic pollution in the ocean requires cooperation from a wide range of stakeholders (e.g., producers, retailers, consumers, researchers), the core regulatory powers of governments are needed for effective solutions (Karasik et al. 2020). Voluntary pledges and commitments alone have been insufficient to manage ocean plastic waste (Borrelle et al. 2020, Cornago, Börkey, and Brown 2021, Lau et al. 2020)—as with many transboundary waste and pollution issues, such as wastewaters degrading basinwide water quality, greenhouse gases causing climate change, air contaminants generating acid rain, and chlorofluorocarbons (CFCs) depleting ozone. A coordinated effort across relevant stages and scales (local, national, and global) is needed to tackle plastic pollution.

The need and opportunity to deploy economic instruments (e.g., the use of taxes and subsidies and extended user responsibilities) *and behavioral interventions* (e.g., promoting the voluntary adoption of pro-environment behavior in societies through non-price and non-regulatory means) to incentivize the most environmentally benign use, recycling, and disposal of plastics and plastic waste (see, e.g., Cornago, Börkey, and Brown 2021).

The opportunities for co-benefits from addressing ocean plastic waste. Reducing plastic waste provides parallel social and environmental benefits for important U.S. priorities, such as equity and environmental justice, climate change emission reduction, sustainable economic growth, and cost reduction (CIEL 2019, Ford et al. 2022, U.S. Department of Energy 2021, UNEP 2021b, World Economic Forum, Ellen MacArthur Foundation, and McKinsey & Company 2016, Zheng and Suh 2019).

> **Recommendation 4:** The United States should create a coherent, comprehensive, and crosscutting federal research and policy strategy that focuses on identifying, implementing, and assessing equitable and effective interventions across the entire plastic life cycle to reduce U.S. contribution of plastic waste to the environment, including the ocean. This strategy should be developed at a high level with a group of experts (or external advisory body) by December 31, 2022, and its implementation assessed by December 31, 2025. Such a strategy would enhance U.S. leadership in creating solutions to global plastic pollution and shaping modern industrial plastic policy.

KEY FRAMEWORKS AND IMPLEMENTATION

No single solution can greatly reduce the flow of plastic waste to the ocean. However, a suite of actions (or "interventions") across all stages of plastics' paths from sources to the ocean could reduce ocean plastic wastes and achieve environmental and social benefits (IRP 2021). Actions to reduce ocean plastic waste at each stage have different effectiveness and costs but together constitute a regional, national, or global strategy for managing plastic wastes in the ocean and the environment (UNEP 2021a). A policy challenge is to organize and implement a portfolio of interventions along this chain of plastic use and management to reduce or eliminate plastic wastes entering the ocean, considering both benefits and costs.

Plastic waste reaching the ocean can be reduced through a range of interventions across the life cycle of plastic waste, from the plastic waste sources to management and release to the ocean (Figure 7.1). Systemic actions in each of these six stages across the plastic life cycle are needed to avoid the current mismatch between (1) sources and production of plastic products and (2) the waste and management systems charged with waste (OECD 2018).

This chapter reviews interventions available and some examples employed to date to prevent and reduce plastic waste from entering the ocean. Interventions managed within a systemic approach can improve outcomes beyond individual interventions.

To reduce plastic waste generation (Stage 3), interventions will be required at the production, material, and product design stages (Stages 1 and 2). These interventions require widespread change in industry standards and practices to make more efficient and equitable use of government and other resources downstream (UNEP 2021a). The federal

Interventions

FIGURE 7.1 Flow diagram of available plastic waste interventions from plastic production to recapture of plastics in the ocean. SOURCE: Modified from Jambeck et al. (2018).

government has a clear opportunity with industry to set goals and requirements to reduce plastic flows from upstream and has laid out some potential innovation paths—for example, the U.S. Department of Energy (2021) Plastics Innovation Challenge Roadmap. At the end of this chapter, Table 7.1 identifies diverse examples for each intervention stage below.

1. *Reduce Plastic Production.* This is the first stage where plastic waste can be affected, by reducing the amount of plastics produced to decrease needs for waste stream management. Of particular interest is reducing production of plastics that are not reusable or practically recyclable.

2. *Innovate Design and Materials.* In this stage, materials and product design innovation can develop substitutes that biodegrade more quickly or are more easily recycled and support use of more reusable products. Furthermore, product design can be changed for items more likely to become waste and leak into the environment through the use of green engineering (Abraham and Nguyen 2003, Anastas and Zimmerman 2003) and green chemistry (Anastas and Warner 1998, Chen et al. 2020, Coish et al. 2018) principles.

3. *Decrease Waste Generation.* Actions in this stage reduce unnecessary plastic wastes, by reducing use of plastic products with short disposable use periods, such as some single-use applications. Such interventions can include product limits and targets for recycling and reuse.

4. *Improve Waste Management.* Actions in this stage improve solid and other waste infrastructure, collection, treatment, and management, including leakage control and accounting. This can include efforts to increase collection of plastics into waste management systems, plastic recycling, and isolation or treatment of remaining plastic wastes to avoid leakage into the environment.

5. *Capture Waste.* Improving waste capture from the environment before or after waste enters the ocean is another class of intervention. This can include re-capturing wastes from ground litter, stormwater, or directly from waters where it accumulates, such as during river or beach cleanups or using retention booms (Figure 7.2). This class of interventions tends to be expensive but is highly visible and often has the most focus.

FIGURE 7.2 A debris retention boom at the Ala Wai Boat Harbor, O'ahu, Hawai'i preventing upstream debris transported via the Ala Wai Canal from entering coastal waters. Image courtesy of Mary J. Donohue.

Environmental capture is sometimes done after plastic wastes enter the open ocean. This strategy is very expensive, inefficient, and impractical because of the vast areas over which waste is dispersed, especially plastic waste that has fragmented over time into very small and widely distributed microplastics.

6. *Minimize at-Sea Disposal.* This category reduces plastic waste discharge into the ocean directly from vessels, point sources, or platforms and includes actions under specific laws and treaties regarding ocean pollution.

Successful implementation of this suite of interventions will require focused resources and funding, as well as attendant monitoring and assessment (as described in Chapter 6), research and development, and public outreach and transparency initiatives (see examples at end of Table 7.1) (Cornago, Börkey, and Brown 2021, UNEP 2021a).

Assessing Interventions—Scale and Cost-Effectiveness

The mix of interventions and actions available to reduce ocean plastic waste constitutes a portfolio within an overall system. If each ocean plastic intervention is managed well, a portfolio of actions will maximize reduction of plastic waste in the ocean for any level of overall cost (IRP 2021). Addressing only one or several categories of interventions without substantially addressing all will reduce overall effectiveness in plastic reduction to the ocean (Biron 2020, Cornago, Börkey, and Brown 2021, Lau et al. 2020).

Actions by larger organizations with the ability to finance, organize, and implement change (e.g., governments and industries) are more likely to have economies of scale in cost and in technical attention to focus on the underlying systems (IRP 2021).

The range of interventions to reduce ocean plastic wastes varies in effectiveness and cost relative to benefits for affected communities and environments. System analysis can help in crafting national, state, and local portfolios of actions, which are more cost-effective and usefully inform policy formulations and discussions.

Participants and Roles

The ubiquity of plastics in the economy and environment is mirrored by the diverse range of institutions and interests involved in the plastic value chain, from plastic production to product manufacture and distribution, disposal, leakage, collection, and recovery or disposal of plastic waste. It is critical to assign roles and responsibilities to those best positioned to address and solve the problem (UNEP 2021a). Multiple interests often need to collaborate for an individual intervention or portfolio of interventions to succeed.

Private-sector groups include raw material feedstock producers; plastic resin producers; plastic processors; designers and creators of plastic products; companies that use plastics in consumer products; and retailers, packagers, and distributors of those products to users ranging from the public to governments. The final stage of the plastic value chain rests with those involved in regulating, financing, and operating systems to control pollution and manage the collection, transportation, treatment, and disposal of plastic wastes. These include landfills, recycling, composting, and incineration facilities as well as facilities to capture and contain leakage to the environment, such as wastewater treatment plants. Governments often take these roles, with private firms carrying out many of these responsibilities.

Private companies have mostly commercial and economic interests in producing and consuming or using plastics and plastic goods, in making plastic material and product design decisions, or in collecting and disposing of plastic wastes. Production and manufacturing firms could intervene in early stages of the value chain and use circular economy principles to reduce the creation of plastic waste in the first instance. They can define clear paths for plastics to end-of-life recovery or management, using green chemistry and green engineering principles (Abraham and Nguyen 2003, Anastas and Zimmerman 2003, Law and Narayan 2022, Zimmerman et al. 2020). These principles can be integrated into products, feedstocks, and manufacturing under expanded definitions of performance that include sustainability (Zimmerman et al. 2020). Such approaches would bring polymer scientists, product designers, environmental engineers, and waste professionals together to design materials and products that reduce the likelihood of leakage and pollution by incentivizing their recovery to retain value and feedstock for future uses (Law and Narayan 2022).

Federal, state, and local governments organize and oversee waste and pollution control operations and infrastructure that are increasingly burdened by plastic waste. U.S. environmental law delegates most of these roles to state and local governments under "cooperative federalism." (Ternes and Fulton 2020[1] and see Appendix C). As currently designed, these systems reduce some externalities but still allow substantial plastic leakage as outlined in Chapters 3 and 4. These include solid waste collection and management systems (including litter collection, landfills, and recycling and composting facilities) as well as treatment and monitoring systems. These are supported by a range of fees and taxes largely at state and local levels, although the federal government funds some infrastructure and targeted prevention and cleanup programs.

National and state governments have critical organizing and motivating roles beyond waste management at all stages, including scientific assessment, monitoring and evaluation, goal and priority setting, expert and cross-sector initiatives, financial incentives, and resources to support change, as well as laws and policies that guide actions by the private sector (Coe, Antonelis, and Moy 2019, UNEP 2021a). Key federal government actors include (1) Congress, which provides statutory authority and fiscal resources; (2) the Executive Branch, which implements statutes and creates executive orders that can stimulate change within the federal system, itself a major consumer; and (3) the Judiciary Branch, which interprets law or gives effect to federal decisions (Ternes and Fulton 2020).

Cost of payment for managing plastic wastes in the environment tends to vary along plastics' paths to the ocean. Each action and its

---

[1]This citation was modified after release of a pre-publication version of the report.

costs affect different consumer and producer groups, as well as different local, regional, and national communities and governments. The current disconnect between plastic formulation and product design and end-of-life management creates significant negative externalities when plastic waste "leakage" creates ocean pollution and inequitable impacts (UNEP 2021b). Consumers, communities, and nongovernmental actors, including philanthropy, are not positioned or resourced to change plastic production and waste management, although they can and do catalyze multisector collaborations, raise awareness, support transparency and equity, and advocate for governmental and private-sector changes. They can also participate in cross-sector partnerships to advance innovation and solutions with government and private firms.

STRATEGIES FROM OTHER COUNTRIES/REGIONS

At the global level, national actions on plastic policy before 2018 focused on interventions largely involving specific plastic products, described in Box 7.1. In 2020, in response to a range of international actions, including UN resolutions regarding plastic pollution, the United Nations Environment Programme (UNEP) issued guidance to assist nations in prioritizing actions to reduce plastic pollution with a more systemic approach, based on a practical understanding of sources of pollution, then matching prioritized "hotspots" (based on data) with appropriate interventions (UNEP 2020). By then, a growing number of G7 and G20 countries had already initiated national "systemic" plans and pressed for coordinated plastic strategies and commitments (see Appendix E). These included the EU (and the United Kingdom), Canada, and China. In October 2021, UNEP released Global Assessment of Marine Litter and Plastic Pollution to inform discussions on additional national and international actions (UNEP 2021a).

Although the United States has a range of laws and policies regarding marine debris and plastic waste (see Chapter 3, Appendix C, and Appendix E), the country has not moved to adopt a national system-wide strategy for reducing plastic waste. The United States did not join Canada, France, Germany, Italy, the United Kingdom, and the EU and numerous nongovernmental groups in signing the 2018 G7 "Plastics Charter," committing to (1) attaining 100% reusable, recyclable, and recoverable plastics by 2030; (2) increasing the recycled content of plastic products to at least 50% by 2030; and (3) recycling and reusing at least 55% of plastic packaging by 2030 and recovering 100% of all plastics by 2040. These commitments underpin the national plastic strategies issued by the EU, United Kingdom, and Canada, described below. In 2019 the United States joined all G20 nations in a voluntary commitment to "reduce additional pollution by marine plastic litter to zero by 2050 through a comprehensive life-cycle approach" but has not yet proposed specific measures to achieve this (G20 2021).

European Union

Recognizing the importance of plastic products to the economy of the EU and the world at large, and plastic pollution's serious harms to the environment and human health, the EU is acting to reduce plastic pollution.

The EU's policy on plastics is embedded in its circular economy plan (European Commission 2021). It intends to transform how plastic products are designed, produced, used, and recycled in the EU, guided by specific rules and targets (European Environment Agency 2021). Some key directions in the EU plastic strategy are (1) improving the economics and quality of recycling by instituting "new rules on packaging to improve the recyclability of plastics and increase demand for recycled plastic content"; (2) curbing plastic waste through a directive banning some single-use products, reducing others, and improving collection and reporting of fishing gear (including through extended producer responsibility [EPR] schemes), as well as rules that restrict use of microplastics in products; (3) driving innovation and investment by increasing financial support, "with an additional €100 million to develop smarter and more recyclable plastics materials"; and (4) working with EU's international partners to "devise global solutions and international standards on plastics."[2]

On marine plastic pollution, Arroyo Schnell et al. (2017) classify the EU's marine plastic pollution policies into three categories:

1. *Plastic production and use impacting the ocean.* Relevant policies involve bans or taxes on plastic items and rely on EU Directive 94/62/EC on packaging and associated waste and its amendment in 2015 (2015/720).
2. *Plastic waste disposal entering the ocean.* Several EU member countries have highlighted their implementation of the International Convention for the Prevention of Pollution from Ships (MARPOL) Convention 73/78. Annex V, in particular, deals with the control and prevention of pollution from garbage from plastic waste and other solid wastes.

---

[2]See https://ec.europa.eu/environment/strategy/plastics-strategy_en.

3. *Plastic waste already in the ocean.* There are policies to reduce the amount of waste already present in the marine environment, including research, monitoring, and cleanup activities.

As a signatory to the Basel Convention, on January 1, 2021, the EU also implemented "new rules banning the export of plastic waste from the EU to non-[Organisation for Economic Co-operation and Development] OECD countries, except for clean plastic waste sent for recycling. Exporting plastic waste from the EU to OECD countries and imports into the EU will

---

BOX 7.1
International Trends on Plastic Policy (as of July 2018)

- **Plastic bag regulations**—127 of 192 countries regulate plastic bags restricting free retail distribution; 27 assess taxes on manufacture and production; 30 charge consumer fees.

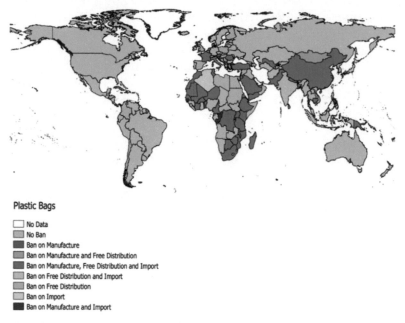

Plastic Bags

☐ No Data
◻ No Ban
■ Ban on Manufacture
▨ Ban on Manufacture and Free Distribution
■ Ban on Manufacture, Free Distribution and Import
▨ Ban on Free Distribution and Import
▨ Ban on Free Distribution
☐ Ban on Import
■ Ban on Manufacture and Import

FIGURE 7.1.1 This map illustrates bans on plastic bags taken by countries around the world. SOURCE: UNEP (2018, Map 1).

- **Product bans or limits**—"27 countries have banned or limited production of specific products (e.g. plates, cups, straws, packaging) and materials (e.g. polystyrene)."

- **Extended producer responsibility (EPR) for plastic bags**—43 countries have included elements of EPR for plastic bags.

- **EPR for single-use plastics**—63 countries mandate EPR for single-use plastics, including deposit-refunds, product take-back, and recycling targets.

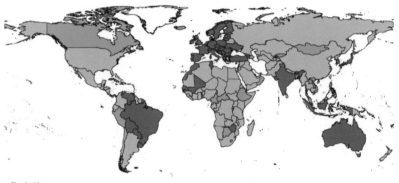

Single Use

☐ No Data
▢ Countries no extended producer responsibility (EPR) or recycling mandates that include single-use plastic items
▢ Countries with recycling mandates that include single-use plastic items but no EPR
■ Countries with EPR for the return, collection, or disposal of single-use plastic items

FIGURE 7.1.2 This map identifies countries that have implemented extended producer responsibility for disposable or single-use plastics. SOURCE: UNEP (2018, Map 8).

- **Microplastics**—Several countries "banned microbeads and the European Union has started a process to restrict the addition of microplastics to consumer and professional use products."

SOURCE: UNEP (2018). For more information, including informative maps, see https://www.unep.org/resources/publication/legal-limits-single-use-plastics-and-microplastics-global-review-national.

also be more strictly controlled."[3] The strategy also provides for periodic evaluation of effectiveness, with some early reports of improvement in recycling of packaging (Hockenos 2021).

Canada

The Canadian Council of Ministers of the Environment Strategy on Zero Plastic Waste (CCME 2019, 2020), adopted in 2018, requires actions along the life cycle of plastics to increase their recovery in the economy. These actions are focused on product design, collection systems, single-use plastics, recycling capacity, and domestic markets for recycled material. This Canada-wide Action Plan on Zero Plastic Waste also includes a Canada-wide Action Plan on Extended Producer Responsibility. On June 9, 2018, Canada also joined France, Italy, the United Kingdom, Germany, and the EU in signing the Ocean Plastics Charter. The Charter's goals include working with industry toward 100% reusable, recyclable, and recoverable plastics by 2030, collaborating with industry and other levels of government to recover 100% of all plastics by 2040.

The Canadian Government has publicly stated that it plans to ban some single-use plastic products, but currently no such legislative bans exist at the federal level. However, a few municipalities are leading the effort on single-use plastic bans. Canada recently adopted a range of legislation and policy statements that will lead to a country-wide ban on single-use plastics by the end of 2021. Six items have been identified for the ban: plastic straws, plastic checkout bags, stir sticks, cutlery, six-pack rings, and foodware made from hard-to-recycle plastics.

Concerning international trade in plastic waste, Canada is a signatory to the Basel Convention, which controls international shipments of most plastic scrap, waste, and waste destined for recycling or disposal (Hagen, LaMotte, and Meng 2021, U.S. EPA 2021j). Canada implements the Basel Convention through the Export and Import of Hazardous Waste and Hazardous Recyclable Material Regulations.

China

In addition to banning plastic waste imports in 2018, China issued national policies on plastic pollution (NDRC and MEE 2020). At the start of 2020, the National Development and Reform Commission (NDRC) and the Ministry of Ecology and Environment (MEE) issued the "Opinions on Further Strengthening the Treatment of Plastic Pollution," which proposed objectives and tasks to phase out certain plastics by 2025 to

---

[3]See https://ec.europa.eu/environment/topics/waste-and-recycling/waste-shipments/plastic-waste-shipments_en.

control plastic pollution (China Government Network 2021). Sequentially, they issued "Notice on Further Strengthening Recent Work of Plastic Pollution Control" and "Notice on Solid Promotion of Plastic Pollution Control" to support their plastic pollution objectives (Guangdong Provincial Development and Reform Commission 2020).

China's guiding principles are to focus on key areas in an "orderly manner," lead through scientific and technological innovation, and foster co-governance (comprising the government, businesses, industry organizations, and the public). China has set goals for 2020, 2022, and 2025 (NDRC and MEE 2020). China's 2020 goal was to become a leader in banning and restricting the "production, sales, and use of some plastic products in some regions and areas." China's 2022 goal is to significantly decrease use of single-use plastic products, promote product substitutes, and increase the proportion of plastic recycled. By 2025, the national objective is to establish a "multi-element" co-governance system along with a management system to address the entire life cycle of plastics, from production to waste. China plans for substitutes for plastic products to be further developed and ready for market, to significantly reduce plastic waste destined for landfills, and to decrease plastic pollution (NDRC and MEE 2020).

In 2008, China attempted to ban ultra-thin plastic bags. The impact was minimal, however, due to insufficient local implementation. Now, China intends to implement and enforce its regulatory measures at the provincial level (Logofet 2021). The NDRC and the MEE recognize that conditions differ in different regions and have stated that local governments should assess local conditions to develop actions and policies for their regions (NDRC and MEE 2020). The central government has requested that provincial governments submit their plans on how to effectively employ the directives for the conditions in their regions (Logofet 2021).

U.S. Federal Action to Date

The United States Federal Strategy for Addressing the Global Issue of Marine Litter, released in October 2020, reflected work as of that date under three main U.S. legal authorities: Marine Debris Act, as amended by the Save our Seas Act of 2018; Resource Conservation and Recovery Act; and Clean Water Act (U.S. EPA, 2020c). It also describes international actions in coordination with other nations under laws on pollution from ships and other ocean activities.

In its September 2021 report to the G20, the United States confirmed that while "it does not have a national action plan specific to marine plastic litter," existing federal laws provide "a comprehensive legal framework to address marine plastic litter," listing, in addition to the three authorities specified in the 2020 Strategy, the Save Our Seas 2.0 Act, the Microbead Free Waters Act, the Toxic Substances Control Act, and the Rivers and Harbors

Appropriation Act (Ministry of Environment, Japan 2021). The 2020 Strategy (U.S. EPA 2020c) and 2021 G20 update confirm that efforts within the United States have focused largely on litter and debris removal, outreach, and monitoring activities, with water pollution and solid waste management and reduction programs delegated to states and local governments.

The U.S. submissions did not include elements adopted by some of the G7 and G20 countries, such as a plan for a national life cycle of plastics intervention strategy or recommend legal, policy, or other changes to reduce production and use of problematic plastics and plastic products as detailed in the previous section. Interventions earlier in the plastic life cycle will be needed to equitably distribute costs and enable interventions to be more effective and cost-efficient (OECD 2018). The only ban on plastic production enacted at the federal level is the congressionally enacted 2015 prohibition on the use and manufacture of rinse-off cosmetic products containing plastic microbeads.

At the same time, states and local jurisdictions have been operating as "policy laboratories" for interventions that have worked elsewhere (Karasik et al. 2020). The need to stem plastic pollution to communities and overburdened waste systems has led some states and local jurisdictions to test new policy tools. Given limited resources and growing public support, states, cities, and municipalities are enacting bans or limits on products (e.g., bags, utensils, and packaging) commonly found in the environment (Karasik et al. 2020). Some states have adopted comprehensive statewide plastic strategies, such as California's 2018 Marine Litter Strategy (co-developed with the National Oceanic and Atmospheric Administration [NOAA]), set for additional updates in 2022 (California Ocean Protection Council and NOAA Marine Debris Program 2018, Wyer 2021). States and local jurisdictions also are adopting policies to redirect recycling and waste management cost from the public sector to producers and generators of plastic and packaging waste. These include EPR laws (e.g., Maine, Oregon) (Martins 2021) and other state policies for various waste types, as noted in Table 7.1 and Appendix C.

A U.S. APPROACH ON PLASTICS

Recent congressional action and federal agency activities, as well as actions adopted by state and local governments, illustrate increasing interest in a more systemic and unified approach to this problem, leading toward a global solution (Save Our Seas 2.0 Act, Reports from NOAA, G20 statements and G7 statement of ministers). The range of federal agencies, programs, and existing legal authorities (illustrated in Appendix C) could be a foundation for an updated U.S. strategy.

As noted, the United States has not yet adopted a systemic federal approach to all six stages of interventions, from production to disposal, though the Save Our Seas 2.0 Act included measures to support research, global cooperation, and infrastructure.[4] Most federal interventions and marine debris strategies within the United States have focused on Stages 3–5, cleanup and local waste management (U.S. EPA, 2020c; Appendixes C and E), which cannot stem leakage to the environment because of the large volume of flow relative to available resources. To reduce U.S. plastic waste generation, interventions will be required in production, material, and product design stages (Stages 1–2). These interventions require widespread changes in industry standards and practices to make the most efficient and equitable use of government and other resources downstream. The federal government has a clear opportunity, along with industry, to set goals and requirements to reduce flows of plastic waste from these upstream stages and has laid out some potential innovation pathways for reducing plastic waste.

NOAA and the U.S. Environmental Protection Agency (U.S. EPA) are two federal agencies with relevant legal authorities and significant expertise in plastic pollution, environmental conservation and protection, and waste management. NOAA's Marine Debris Program was formed when most attention was directed toward plastic and other waste in the ocean and on shorelines from marine-based sources and focused attention on abandoned and lost fishing gear as well as ship-based plastic waste (National Research Council 2009, U.S. Commission on Ocean Policy 2004), but the program has been strengthened multiple times to address new challenges and improve the following: government coordination, including around enforcement of existing laws; public outreach and education; partnerships; monitoring and identification; and research. NOAA has also been a leader internationally, having hosted or co-hosted six International Marine Debris Conferences, including in Hawaii (1984) and San Diego, California (2018). These meetings were an important forum for marine debris researchers, managers, policy makers, and others interested in marine debris. The United States and the Republic of Korea have announced plans for a seventh International Marine Debris Conference in Busan, Republic of Korea in September 2022.

NOAA has made progress on these efforts, building scientific and operational expertise, widespread trust, and strong partnerships within existing resources and authorities. This role as a trusted science-based leader and partner on the problem is essential to the success of any federal effort. In addition, NOAA leads the Interagency Marine Debris Coordinating Committee (IMDCC), under which it coordinates with many federal agencies with programs and resources to bear on the plastic waste problem.

Reducing land-based sources of waste and pollution that enter the U.S. environment, including federal inland and offshore waters, is assigned to

---

[4]See https://www.congress.gov/bill/116th-congress/senate-bill/1982/text.

U.S. EPA, with roles for NOAA, the U.S. Coast Guard, and other agencies (Appendix C). As described in Chapter 3, U.S. EPA's existing environmental authorities, while broad, operate within a federal and state regulatory context. Their water and air pollution prevention and solid and hazardous waste management authorities, largely implemented at local and state levels, are grounded in a historical focus on hazardous waste and chemical pollutants and are not specifically designed to address plastic waste problems. However, U.S. EPA's expertise on strategies for pollution and waste control and human and environmental health risk reduction give the agency a strong opportunity to use its experience in designing critical interventions.

Although the United States is strong in solid waste management compared to other countries, plastic solid waste is primarily landfilled despite major efficiencies and benefits to be gained by interventions in Stages 1–3 to reduce plastic waste and divert plastic waste to other managed fates (recycling, composting, reuse). It will be important to use a range of federal interventions across Stages 1–6 to reduce plastic waste "leakage" into the environment and ocean. The talent of federal agencies and many others will be needed to address gaps in plastic waste source reduction and building the infrastructure and systems to support plastic reduction, reuse, recycling, or composting (see Appendix C).

This report does not review the state of knowledge on impacts of plastic waste to humans and the environment, but such an assessment could be an important part of developing a national strategy to inform necessary and priority actions across intervention stages. For example, the United States could consider whether it is appropriate to regulate plastic waste as a pollutant or hazardous material based on such an assessment.

Finally, the federal research and monitoring enterprise is not resourced or organized to bring the needed science and assessments to bear on research priorities relating to the entire life cycle and scope of plastics, or key intervention points identified in this and other expert reports (UNEP 2021a). NOAA has led the federal monitoring and assessment effort and, along with U.S. EPA, conducted research and provided small-scale external research grants. However, as noted in Chapter 6, most research on the extent of plastics in the ocean and the natural environment has been undertaken by scientists outside of federal agencies, funded through both federal and non-federal sources. Emerging federal research and development, initiatives, and public–private collaborations may support more innovation on a range of topics, including materials design (see Table 7.1, Stage 2), but these efforts are in the early stages.

Monitoring and assessments on plastic pollution will require more federal coordination, resources, and attention. Ensuring that the work is strategic and targeted to support top interventions would benefit from being organized at a higher level of government, such as has been done for many transboundary environmental challenges (e.g., climate change, transboundary pollutants such as CFCs and oil). Models for such high-level federal science coordination exist, such as under the National Science and Technology Council, the U.S. Global Change Research Program, the U.S. Ocean Policy, and coordination mechanisms such as the IMDCC and interagency ocean observing system committees (Appendix C).

The challenges of implementing a coherent U.S. portfolio of effective system-wide interventions can be targeted and overcome by a new national strategy and implementation plan that builds on existing legal authorities and agency efforts, adopts new models being tested by others, and fills gaps identified above, in Table 7.1, and in Appendix C. Such a system can (1) provide a clear policy and legal framework and goals for reducing plastic waste in the ocean, (2) create economic incentives for improved plastic manufacturing and reduction through reuse and recycling, (3) reduce plastic "leaks" in U.S. waste management and pollution control systems, and (4) address funding gaps and reverse inequitable cost burdens.

An updated U.S. strategy should take a systemic view and better organize actions across the range of federal agencies and programs (Appendix C), as well as state, tribal, and local governments, and other important industry, philanthropy, science, and civil society involvement. It could reflect new information and models for action, such as those being developed and tested by U.S. states and other countries, described above and in Table 7.1. The differences between the current U.S. approach and those being implemented elsewhere, as well as analyses of their effectiveness (see Cornago, Börkey, and Brown 2021, UNEP 2021a), could provide priority areas for evaluation with experts.

Using these resources, the United States could update its policy, goals, and legal framework to reduce the U.S. contribution to global ocean plastic waste and assess this progress. High-level goals could be tailored to identify and address gaps in the U.S. system and unite federal efforts around specific coordinated interventions.

Creating a framework for a system of interventions can align the United States with a global approach (Appendix E). Action could focus on those interventions suited to address the problem and reduce barriers to action. Moreover, U.S. leadership would help position the nation to shape and influence global activities in plastic production, formulation, design, innovation, and waste reduction. This, in turn, can create innovation and economic opportunities that reduce negative economic externalities.

TABLE 7.1 Non-comprehensive Table of Intervention Options Along Plastic Waste's Path to the Ocean

Intervention Category	Types of Interventions
1. Reduce Plastic Production	
Production or manufacturing restrictions and limits	National goals and strategies to cap or reduce virgin plastic production
	Reductions in plastic production (as carbon equivalents) as part of global, U.S., and state greenhouse gas emissions goals
	Moratorium on new petrochemical plants and capacity to reduce production from fossil feedstocks
2. Innovate Material and Product Design	
Enforceable product standards for manufacturers	Timebound targets and limits on plastic content of specific products and packaging
	End-of-life material and design specifications (simplification) for some products, packaging to facilitate reuse, recycling
Voluntary commitments and collaborations for innovative material and product design	Government-sponsored research and development collaborations, incentives, and roadmaps (see also "Other Activities" below)
	Promote industry-wide innovation, standards, collaboration, and regulation by constraining the types of resins used in some applications to maximize value and recyclability
	Streamline and standardize design to limit variability in packaging
	End-of-life material and design specifications (simplification) for some products, packaging to facilitate reuse, recycling
	Encourage following the principles of green engineering and green chemistry

Implementers	Specific Illustrative Examples
National, state, and tribal governments and industry standards	European Union (EU) Circular Economy Action Plan, March 11, 2020, and EU Directives 2018/850 and 2018/851 (landfill limits and recycling targets)
National and state governments, standards organizations Industry (standards and systems)	Minimum recycled content requirements (California bottle recycled content law [Keller and Heckman 2020]; Washington state and Connecticut [LaMotte et al. 2021])
	EU Directive 2018/852 (minimum 55% recycled content in plastic packaging by 2030)
	Prohibitions on sale of packaging with some plastics, such as polystyrene (e.g., Washington State SB5022, enacted 2021 [Quinn 2021])
Industry, government, academia, nongovernmental (scientific, funding, environmental) organizations, global standards organizations	U.S. Plastics Pact[a]
	Precompetitive and open innovation collaborations within and across industry sectors (e.g., Ellen MacArthur Circular Economy 100 Group [Kleine Jäger and Piscicelli 2021])
	SOS 2.0 Genius Prize for Save our Seas Innovations (Department of Commerce and new Marine Debris Foundation)

continued

TABLE 7.1 Continued

Intervention Category	Types of Interventions
Standards for labeling and marketing	Restrict use of chasing arrows symbol on products which lack broad, functional recycling infrastructure (e.g., can be collected, sorted, cleaned, and economically reprocessed) in place in the United States
	Restrict chasing arrows symbol to items following material standards for that product or material
	Create enforceable feedstock, performance, and labeling standards for "biodegradable," "compostable," "biobased" products, to prevent consumer confusion and potential "greenwashing"
	Publicly available assessments of and reports on recycling efficacy (markets for recycled materials and fate of items collected in recycling process)
3. Decrease Waste Generation	
Plastic product bans (and substitutes)	Ban specific products based upon criteria such as potential for loss to the environment, toxicity, and necessity of use
Mandatory procurement rules favoring reusable products	Procurement rules to replace single-use items with reusable goods
Reduce loss of pre-production pellets that become waste	Reduce pellet losses and wastes
Fiscal tools (fees, taxes, incentives)	Fee on purchase of specific items at point-of-sale to disincentivize their use (e.g., thin film shopping bags)
Deposit return systems	Systems that use a deposit to incentivize return or reuse of the packaging or product
Extended producer requirements (EPR) (end-of-life management)	Place legal or fiscal responsibility on producers for management and disposal of plastic waste. EPR campaigns often rely on government to set and enforce standards even though responsibility is placed upon companies.
	Laws and policies that enable life-cycle management such as EPR, take back schemes that meet specific targets for waste diversion and recycling
	Require recycling rates for products (e.g., beverage bottles). If rates are not met, then fees are charged.

Implementers	Specific Illustrative Examples
National, state, and tribal governments; consumers and civil society	U.S. Federal Trade Commission Green Guides for Environmental Marketing Claims
	CA SB 343 (restricts use of the chasing arrows symbol to only those plastic products that are truly recyclable in California); CA AB 1201 (restricts manufacturers from making the compostable claim unless the product meets specific compostability criteria)
	Nongovernmental and governmental reports (e.g., Greenpeace 2020, U.S GAO 2020)
National, state, local, and tribal governments	EU Directive 2019/904 (Single-Use Product Ban), effective 2021
	Various U.S. state and local bans on single-use products (bags, straws, food service items); See Box 7.1 and Appendix C
National, state, and tribal governments	Canada 2018 Strategy: Zero plastic waste (Government of Canada 2021)
Private-sector companies, nongovernmental institutions	
National and state governments; industry	2007 California law (AB 258) on pre-production plastic source control[b]
National, state, municipal, and tribal governments, and consumers	U.S. state and municipal plastic bag laws
	U.S. state bottle return laws (see Appendix C)
	Norway tax on plastic producers, forgiven if recycling tops 95% (now 97% bottles are recycled; 92% can be reused) (Steffen 2020)
National, state and local, and tribal governments	Maine and Oregon packaging EPR laws (2021) and other state EPR laws
Industry funded/ government oversight	British Columbia EPR law (85% recovery rate; Paben 2021)
	Many plastic and non-plastic examples in states (e.g., paint, mattresses)[c]
	U.S. EPR requirements for e-waste and pharmaceuticals
	EU and Norway EPR legislation

continued

TABLE 7.1 Continued

Intervention Category	Types of Interventions
Reusable and refillable systems	Investment in affordable and convenient reuse/refill systems to reduce single-use packaging
	Fund programs to promote reuse/refill systems

4. Improve Waste Management (Prevent or Reduce Disposal/Discharge)	
Disposal, collection, and recycling improvements	Infrastructure for source separation, industrial composting, recycling (including beyond mechanical)
	Recycling collection and reuse targets and incentives (e.g., bottle bills, deposit refund schemes)
	Place and maintain receptacles in plastic "hotspot" or high-traffic areas
	Research and development investment in new methods of depolymerizing plastic waste to promote material/chemical recovery
Plastic waste export/import controls	Limit, ban, or voluntarily eliminate plastic waste exports and imports to incentivize waste reduction
Treatment improvements to remove plastic waste from discharges	Wastewater treatment standards to remove microplastics and microfibers
	Products to prevent microfiber releases from equipment (e.g., washing and industrial machines)
National Pollutant Discharge Elimination System, stormwater limits and treatment	Stormwater discharge regulations for plastics
	Green infrastructure to filter stormwater
Ocean/river discharge limits	Establish regulatory limits on macroplastic or microplastic waste in ocean and river discharges

Implementers	Specific Illustrative Examples
National, state, and tribal governments Investment through Small Business Innovation Research, government funding, private funders	CA laws: (1) AB 962 allows beverage producers to sanitize and refill intact glass bottles; (2) CA AB 619^d—amends health laws to allow consumers to bring containers for restaurants to fill for to-go. Business examples: Algramo, Loop
National, state, tribal, and local governments	Infrastructure grants under Save Our Seas 2.0 Act and related legislation (Clean Water Act and Safe Drinking Water Act; see Appendix C) State bottle bills (e.g., CA AB 962^e requires the creation of a returnable bottle system in California by January 1, 2024) Cigarette butt bins Lidded trash cans U.S. Department of Energy investments (e.g., Energy. gov 2020); industry initiatives and multiparty alliances; see also research and development, below.
National, state, and tribal governments; private sector	None at federal level (not signatory to Basel Convention) CA AB 881 prevents municipalities from counting plastic waste exports as "recycled" Private industry voluntary commitments (Waste Management, Republic Services) China 2018 Import Ban Basel Convention 2019 amendments (require prior informed consent for exports of hazardous plastic waste and most non-hazardous plastic waste)
Government, private sector	California requires plastic waste removal from industrial and municipal discharge
National, state, and tribal governments	California, Hawaii Trash total maximum daily loads to address plastic waste in stormwater Nonpoint source permit requirements (facility specific, per U.S. Environmental Protection Agency guidance)
National, state, and tribal governments	California zero discharge goal for trash (including plastics) by 2030

continued

TABLE 7.1 Continued

Intervention Category	Types of Interventions
5. Capture Waste (to Remove Plastic Waste from the Environment)	
Remove wastes from waterways	Beach, river, and inland waterway cleanups
	Trash capture devices in waterways
Remove wastes from ocean wildlife and habitats	Ghost net removal; fishing gear return incentives; animal and coral disentanglement
Remove plastic waste from localized hotspots	Tire wear particle capture device for roadways
	Land-based cleanups
	Research to identify plastic waste hotspots
6. Minimize Ocean Disposal	
Increase enforcement for at-sea disposal	Increase enforcement of dumping and disposal of trash
Reduce at-sea abandonment or discard of fishing gear	Establish solid waste disposal infrastructure for end-of-life fishing nets and gear
	Create incentives for land-based, e.g., dockside, disposal of end-of-life fishing nets, gear, and trash
	Establish identification/tagging for deployed active and passive fishing nets and pots
Other Activities (to Support Above Interventions)	
Information/data collection	Coordinated tracking and monitoring systems
	Community-based monitoring
	National and state economic data, field data and studies
	Mandatory annual reports on plastic use inventories of public companies and government institutions
	Require plastic producers to report plastic production on carbon equivalents

Implementers	Specific Illustrative Examples
Municipal governments, community groups	International Coastal Cleanup/Ocean Conservancy
	Mr. Trash Wheel, trash booms, etc.
National, state, local, and tribal governments; local, industry, and nonprofit groups	Derelict crab pot removal
	Global Ghost Gear Initiative/Ocean Conservancy
	Sustainable Coastlines Hawaii/State of Hawaii Marine Debris Rapid Response Ghost Net Removal Program and marine litter removal
	U.S. National Oceanic and Atmospheric Administration National Marine Fisheries Service Pacific Islands Fishery Science Center ghost net removal, protected species disentanglement
	U.S. National Oceanic and Atmospheric Administration Marine Debris Program funded community-based marine debris removal projects
	Hawaii Pacific University Center for Marine Debris Research ghost net removal in state of Hawaii
	The Northwest Straits Foundation ghost net and derelict crab pot removal in Puget Sound
State, local, and tribal governments	Cleanup efforts
Academia, nongovernmental organizations, agencies	
Global treaty organizations; national, state, local, and tribal governments	MARPOL VI; Ocean Dumping Act implementation measures
	EU Directive 2019/904 provides for EPR and proper disposal of fishing gear made of plastics
	Various national and state fishing gear marking requirements (e.g., Marine Management Organisation 2016, Ocean Outcomes 2020)
National, state, local, and tribal governments; industry	Marine Debris Monitoring and Assessment Project, U.S. National Water Quality Monitoring Council/Marine Debris Tracker, International Coastal Cleanup/CleanSwell, regional and local activities
	Transparency reporting: (1) Shareholder and investor initiatives (e.g., "As You Sow"), (2) Public reporting (e.g., "Plastic Waste Makers Index," Minderoo Foundation)

continued

TABLE 7.1 Continued

Intervention Category	Types of Interventions
Research and development	Methods to deliver products without packaging
	Industrially compostable and home compostable polymers, films, and adhesives
	Product design that maximizes circularity and recyclability
	Circular materials management and leakage characterization to inform upstream interventions
	Intersectional and interdisciplinary research to prevent litter and illegal dumping
Education and outreach	Professional outreach, co-production of knowledge to inform solutions at local and regional scales
	Outreach on efficacy of plastic recycling, labeling, and engaging public in solutions
	Media, school materials, aquaria, and museums including information on ocean plastics
	Public behavior-change campaigns
	Community outreach to identify and address local barriers to prevent litter, illegal dumping

[a] See https://usplasticspact.org/.
[b] See https://www.waterboards.ca.gov/water_issues/programs/stormwater/plasticdebris.shtml.
[c] See https://www.productstewardship.us/page/State_EPR_Laws_Map.
[d] See https://leginfo.legislature.ca.gov/faces/billTextClient.xhtml?bill_id=201920200AB619.
[e] See https://leginfo.legislature.ca.gov/faces/billNavClient.xhtml?bill_id=202120220AB962.
[f] See https://acwi.gov/monitoring/network/index.html.
[g] See https://pledge.ourhands.org/.
[h] See www.trashshouldntsplash.org.

Implementers	Specific Illustrative Examples
	REMADE Institute
	U.S. Department of Energy Plastic Innovation Roadmap
	National Science Foundation (NSF) Convergent Accelerator program and NSF Grand Challenges grants
	Ellen MacArthur Foundation Plastics Pacts; American Chemistry Council Roadmap to Reuse
	Trash Free Seas Alliance; Global Plastics Alliance and related industry investments and partnerships
	New Materials Institute Center for Bioplastics and Biocomposites
All	National Oceanic and Atmospheric Administration National Sea Grant College Program
	Nongovernmental organization and governmental reports, data, outreach
	U.S. coastal and inland aquarium (Aquarium Conservation Partnership) outreach campaigns on single-use plastics: "In Our Hands" (2017)[g] and (2) "First Step" on straws (2018)
	Trash Shouldn't Splash[h]
	Space Apps Challenge, e.g., 2021 Challenge–Leveraging AI/ML for Plastic Marine Debris

References

Abbott, J. K., and U. R. Sumaila. 2019. "Reducing marine plastic pollution: Policy insights from economics." *Review of Environmental Economics and Policy* 13 (2):327-336. doi: 10.1093/reep/rez007.

Abraham, M. A., and N. Nguyen. 2003. "'Green engineering: Defining the principles'—Results from the Sandestin Conference." *Environmental Progress* 22 (4):233-236. doi: 10.1002/ep.670220410.

Abt Associates. 2019. *The Effects of Marine Debris on Beach Recreation and Regional Economies in Four Coastal Communities: A Regional Pilot Study.* Silver Spring, MD: National Oceanic and Atmospheric Administration.

Acuña-Ruz, T., D. Uribe, R. Taylor, L. Amézquita, M. C. Guzmán, J. Merrill, P. Martínez, L. Voisin, and C. Mattar. 2018. "Anthropogenic marine debris over beaches: Spectral characterization for remote sensing applications." *Remote Sens Environ* 217:309-322.

Adomat, Y., and T. Grischek. 2021. "Sampling and processing methods of microplastics in river sediments - A review." *Sci Total Environ* 758:143691. doi: 10.1016/j.scitotenv.2020.143691.

Agamuthu, P., S. B. Mehran, A. Norkhairah, and A. Norkhairiyah. 2019. "Marine debris: A review of impacts and global initiatives." *Waste Manag Res* 37 (10):987-1002. doi: 10.1177/0734242X19845041.

Alexander, M. 1975. "Environmental and microbiological problems arising from recalcitrant molecules." *Microb Ecol* 2 (1):17-27. doi: 10.1007/BF02010378.

Alimi, O. S., J. Farner Budarz, L. M. Hernandez, and N. Tufenkji. 2018. "Microplastics and nanoplastics in aquatic environments: Aggregation, deposition, and enhanced contaminant transport." *Environ Sci Technol* 52 (4):1704-1724. doi: 10.1021/acs.est.7b05559.

Allen, H. 2021. *Microplastics Pollution—The Legacy and Sentinel of Mismanaged Plastic Waste.* U.S. EPA.

Allen, R., D. Jarvis, S. Sayer, and C. Mills. 2012. "Entanglement of grey seals *Halichoerus grypus* at a haul out site in Cornwall, UK." *Mar Pollut Bull* 64 (12):2815-2819. doi: 10.1016/j.marpolbul.2012.09.005.

Allen, S., D. Allen, K. Moss, G. Le Roux, V. R. Phoenix, and J. E. Sonke. 2020. "Examination of the ocean as a source for atmospheric microplastics." *PLoS ONE* 15 (5). doi: 10.1371/journal.pone.0232746.

Allen, S., D. Allen, V. R. Phoenix, G. Le Roux, P. Durántez Jiménez, A. Simonneau, S. Binet, and D. Galop. 2019. "Atmospheric transport and deposition of microplastics in a remote mountain catchment." *Nat Geosci* 12 (5):339-344. doi: 10.1038/s41561-019-0335-5.

Alzugaray, L., M. Di Martino, L. Beltramino, V. J. Rowntree, M. Sironi, and M. M. Uhart. 2020. "Anthropogenic debris in the digestive tract of a southern right whale (*Eubalaena australis*) stranded in Golfo Nuevo, Argentina." *Mar Pollut Bull* 161 (Pt A):111738. doi: 10.1016/j.marpolbul.2020.111738.

Amaral-Zettler, L. A., C. L. Dupont, E. R. Zettler, B. Slikas, D. Kaul, and T. J. Mincer. 2016. "The Plastisphere 'Microbiome.'" Paper presented at American Geophysical Union Ocean Sciences Meeting. 2016:HI41A-03.

Amaral-Zettler, L. A., E. R. Zettler, and T. J. Mincer. 2020. "Ecology of the plastisphere." *Nat Rev Microbiol* 18 (3):139-151. doi: 10.1038/s41579-019-0308-0.

Amato, D. W., R. B. Whittier, H. Dulai, and C. M. Smith. 2020. "Algal bioassays detect modeled loading of wastewater-derived nitrogen in coastal waters of O'AHU, HAWAI'I." *Mar Pollut Bull* 150. doi: 10.1016/j.marpolbul.2019.110668.

American Chemical Society. 2021. "12 Principles of Green Engineering." https://www.acs.org/content/acs/en/greenchemistry/principles/12-design-principles-of-green-engineering.html.

American Chemistry Council. 2021a. "Background and Methodology for Statistical Reports." https://plastics.americanchemistry.com/background-and-methodology-for-statistical-reports/.

American Chemistry Council. 2021b. "Studies and Information: Resin Report Subscriptions." https://plastics.americanchemistry.com/resin-report-subscriptions/.

Ammendolia, J., J. Saturno, A. L. Brooks, S. Jacobs, and J. R. Jambeck. 2021. "An emerging source of plastic pollution: Environmental presence of plastic personal protective equipment (PPE) debris related to COVID-19 in a metropolitan city." *Environ Pollut* 269:116160. doi: 10.1016/j.envpol.2020.116160.

Anastas, P., and J. C. Warner. 1998. *Green Chemistry: Theory and Practice*. New York: Oxford University Press.

Anastas, P. T., and J. B. Zimmerman. 2003. "Design through the 12 principles of green engineering." *Environ Sci Technol* 37 (5):94A-101A. doi: 10.1021/es032373g.

Anbumani, S., and P. Kakkar. 2018. "Ecotoxicological effects of microplastics on biota: A review." *Environ Sci Pollut Res* 25 (15):14373-14396. doi: 10.1007/s11356-018-1999-x.

Andrady, A. L. 2011. "Microplastics in the marine environment." *Mar Pollut Bull* 62 (8): 1596-1605.

Andrady, A. L., and M. A. Neal. 2009. "Applications and societal benefits of plastics." *Philos Trans R Soc Lond B Biol Sci* 364 (1526):1977-1984. doi: 10.1098/rstb.2008.0304.

Araujo, C. F., M. M. Nolasco, A. M. P. Ribeiro, and P. J. A. Ribeiro-Claro. 2018. "Identification of microplastics using Raman spectroscopy: Latest developments and future prospects." *Water Res* 142:426-440. doi: 10.1016/j.watres.2018.05.060.

Arias-Andres, M., U. Klümper, K. Rojas-Jimenez, and H. P. Grossart. 2018. "Microplastic pollution increases gene exchange in aquatic ecosystems." *Environ Pollut* 237:253-261. doi: 10.1016/j.envpol.2018.02.058.

Arii, M., M. Koiwa, and Y. Aoki. 2014. "Applicability of SAR to marine debris surveillance after the Great East Japan Earthquake." *IEEE J Sel Top Appl Earth Obs Remote Sens* 7 (5): 1729-1744.

Arp, H. P. H., D. Kühnel, C. Rummel, M. Macleod, A. Potthoff, S. Reichelt, E. Rojo-Nieto, M. Schmitt-Jansen, J. Sonnenberg, E. Toorman, and A. Jahnke. 2021. "Weathering plastics as a planetary boundary threat: Exposure, fate, and hazards." *Environ Sci Technol* 55 (11):7246-7255. doi: 10.1021/acs.est.1c01512.

Arroyo Schnell, A., N. Klein, E. Gómez Girón, and J. Sousa. 2017. *National Marine Plastic Litter Policies in European Union Member States: An Overview*. Brussels: International Union for Conservation of Nature.

Bai, Y., and J. Givens. 2021. "Ecologically unequal exchange of plastic waste?" *J World-Sys Res* 27 (1):265-287. doi: 10.5195/jwsr.2021.1026.

Balas, C. E., A. T. Williams, S. L. Simmons, and A. Ergin. 2001. "A statistical riverine litter propagation model." *Mar Pollut Bull* 42 (11):1169-1176. doi: 10.1016/s0025-326x(01) 00133-3.

Baldwin, A. K., S. R. Corsi, and S. A. Mason. 2016. "Plastic debris in 29 Great Lakes tributaries: Relations to watershed attributes and hydrology." *Environ Sci Technol* 50 (19):10377-10385. doi: 10.1021/acs.est.6b02917.

Bank, M. S., and S. V. Hansson. 2019. "The plastic cycle: A novel and holistic paradigm for the Anthropocene." *Environ Sci Technol* 53 (13):7177-7179. doi: 10.1021/acs.est.9b02942.

Bao, Z., J. Sha, X. Li, T. Hanchiso, and E. Shifaw. 2018. "Monitoring of beach litter by automatic interpretation of unmanned aerial vehicle images using the segmentation threshold method." *Mar Pollut Bull* 137:388-398. doi: 10.1016/j.marpolbul.2018.08.009.

Barnes, D. K. A. 2002. "Biodiversity: Invasions by marine life on plastic debris." *Nature* 416 (6883):808-809. doi: 10.1038/416808a.

Barnes, D. K., F. Galgani, R. C. Thompson, and M. Barlaz. 2009. "Accumulation and fragmentation of plastic debris in global environments." *Philos Trans R Soc Lond B Biol Sci* 364 (1526):1985-1998. doi: 10.1098/rstb.2008.0205.

Barrett, J., Z. Chase, J. Zhang, M. M. Banaszak Holl, K. Willis, A. Williams, B. D. Hardesty, and C. Wilcox. 2020. "Microplastic pollution in deep-sea sediments from the Great Australian Bight." *Front Mar Sci* 7. doi: 10.3389/fmars.2020.576170.

Barrows, A. P. W., K. S. Christiansen, E. T. Bode, and T. J. Hoellein. 2018. "A watershed-scale, citizen science approach to quantifying microplastic concentration in a mixed land-use river." *Water Res* 147:382-392. doi: 10.1016/j.watres.2018.10.013.

Baulch, S., and C. Perry. 2014. "Evaluating the impacts of marine debris on cetaceans." *Mar Pollut Bull* 80 (1-2):210-221. doi: 10.1016/j.marpolbul.2013.12.050.

Baumol, W. J., and W. E. Oates. 1988. *The Theory of Environmental Policy*. Cambridge, UK: Cambridge University Press.

Beaumont, N. J., M. Aanesen, M. C. Austen, T. Börger, J. R. Clark, M. Cole, T. Hooper, P. K. Lindeque, C. Pascoe, and K. J. Wyles. 2019. "Global ecological, social and economic impacts of marine plastic." *Mar Pollut Bull* 142:189-195. doi: 10.1016/j.marpolbul. 2019.03.022.

Beck, C. A., and N. B. Barros. 1991. "The impact of debris on the Florida manatee." *Mar Pollut Bull* 22 (10):508-510. doi: 10.1016/0025-326X(91)90406-I.

Ben-Haim, Y., F. L. Thompson, C. C. Thompson, M. C. Cnockaert, B. Hoste, J. Swings, and E. Rosenberg. 2003. "*Vibrio coralliilyticus* sp. nov., a temperature-dependent pathogen of the coral *Pocillopora damicornis*." *Int J Syst Evol Microbiol* 53 (Pt 1):309-315. doi: 10.1099/ ijs.0.02402-0.

Besseling, E., E. M. Foekema, J. A. van Franeker, M. F. Leopold, S. Kühn, E. L. Bravo Rebolledo, E. Heße, L. Mielke, J. Ijzer, P. Kamminga, and A. A. Koelmans. 2015. "Microplastic in a macro filter feeder: Humpback whale *Megaptera novaeangliae*." *Mar Pollut Bull* 95 (1): 248-252. doi: 10.1016/j.marpolbul.2015.04.007.

Biermann, L., D. Clewley, V. Martinez-Vicente, and K. Topouzelis. 2020. "Finding plastic patches in coastal waters using optical satellite data." *Sci Rep* 10 (5364). doi: 10.1038/ s41598-020-62298-z.

Bilkovic, D. M., K. J. Havens, D. M. Stanhope, and K. T. Angstadt. 2012. "Use of fully biodegradable panels to reduce derelict pot threats to marine fauna." *Conserv Biol* 26 (6):957-966. doi: 10.1111/j.1523-1739.2012.01939.x.

Biron, M. 2020. "Plastics sustainability: Drivers and obstacles." In *A Practical Guide to Plastics Sustainability*, edited by M. Biron, 557-593. Norwich: William Andrew Publishing.

Blettler, M. C. M., and K. M. Wantzen. 2019. "Threats underestimated in freshwater plastic pollution: Mini-review." *Water Air Soil Pollut* 230 (7). doi: 10.1007/s11270-019-4220-z.

Boland, R. C., and M. J. Donohue. 2003. "Marine debris accumulation in the nearshore marine habitat of the endangered Hawaiian monk seal, *Monachus schauinslandi* 1999–2001." *Mar Pollut Bull* 46 (11):1385-1394.

Bond, K., H. Benham, E. Vaughan, and L. Chau. 2020. "The Future's Not in Plastics: Why Plastics Demand Won't Rescue the Oil Sector." Analyst Note. Carbon Tracker.

Boren, L. J., M. Morrissey, C. G. Muller, and N. J. Gemmell. 2006. "Entanglement of New Zealand fur seals in man-made debris at Kaikoura, New Zealand." *Mar Pollut Bull* 52 (4):442-446. doi: 10.1016/j.marpolbul.2005.12.003.

Borrelle, S. B., J. Ringma, K. L. Law, C. C. Monnahan, L. Lebreton, A. McGivern, E. Murphy, J. Jambeck, G. H. Leonard, M. A. Hilleary, M. Eriksen, H. P. Possingham, H. De Frond, L. R. Gerber, B. Polidoro, A. Tahir, M. Bernard, N. Mallos, M. Barnes, and C. M. Rochman. 2020. "Predicted growth in plastic waste exceeds efforts to mitigate plastic pollution." *Science* 369 (6510):1515-1518. doi: 10.1126/science.aba3656.

Botterell, Z. L. R., N. Beaumont, T. Dorrington, M. Steinke, R. C. Thompson, and P. K. Lindeque. 2019. "Bioavailability and effects of microplastics on marine zooplankton: A review." *Environ Pollut* 245:98-110. doi: 10.1016/j.envpol.2018.10.065.

Bourne, W. R. P., and M. J. Imber. 1982. "Plastic pellets collected by a prion on Gough Island, Central South America." *Mar Pollut Bull* 13 (1):20-21.

Brahney, J., N. Mahowald, M. Prank, G. Cornwell, Z. Klimont, H. Matsui, and K. A. Prather. 2021. "Constraining the atmospheric limb of the plastic cycle." *Proc Natl Acad Sci USA* 118 (16). doi: 10.1073/pnas.2020719118.

Brander, S., E. Hoh, K. Unice, A.-M. Cook, R. Holleman, C. M. Rochman, and J. Thayer. 2021. *Microplastic Pollution in California: A Precautionary Framework and Scientific Guidance to Assess and Address Risk to the Marine Environment*. Sacramento: California Ocean Science Trust.

Bravo Rebolledo, E. L., J. A. van Franeker, O. E. Jansen, and S. M. Brasseur. 2013. "Plastic ingestion by harbour seals (*Phoca vitulina*) in The Netherlands." *Mar Pollut Bull* 67 (1-2):200-202. doi: 10.1016/j.marpolbul.2012.11.035.

British Plastics Federation. 2019. "Oil Consumption." https://www.bpf.co.uk/press/Oil_Consumption.

Britt, P. F., G. W. Coates, K. I. Winey, J. Byers, E. Chen, B. Coughlin, C. Ellison, J. Garcia, A. Goldman, J. Guzman, J. Hartwig, B. Helms, G. Huber, C. Jenks, J. Martin, M. McCann, S. Miller, H. O'Neill, A. Sadow, S. Scott, L. Sita, D. Vlachos, and R. Waymouth. 2019. Basic Energy Sciences Roundtable on Chemical Upcycling of Polymers. Department of Energy.

Brock, J. 2020. "The Plastics Pandemic: COVID-19 Trashed the Recycling Dream." *Reuters*. https://www.reuters.com/investigates/special-report/health-coronavirus-plastic-recycling/.

Brooks, A. 2021. "From the Ground Up: Measurement, Review, and Evaluation of Plastic Waste Management at Varying Landscape Scales." PhD Dissertation, University of Georgia.

Brooks, A. L., S. Wang, and J. R. Jambeck. 2018. "The Chinese import ban and its impact on global plastic waste trade." *Sci Adv* 4 (6):eaat0131. doi: 10.1126/sciadv.aat0131.

Browne, M. A., M. G. Chapman, R. C. Thompson, L. A. Amaral Zettler, J. Jambeck, and N. J. Mallos. 2015. "Spatial and temporal patterns of stranded intertidal marine debris: Is there a picture of global change?" *Environ Sci Technol* 49 (12):7082-7094. doi: 10.1021/es5060572.

Bryant, J. A., T. M. Clemente, D. A. Viviani, A. A. Fong, K. A. Thomas, P. Kemp, D. M. Karl, A. E. White, and E. F. DeLong. 2016. "Diversity and activity of communities inhabiting plastic debris in the North Pacific Gyre." *mSystems* 1 (3). doi: 10.1128/mSystems.00024-16.

Bucci, K., M. Tulio, and C. M. Rochman. 2020. "What is known and unknown about the effects of plastic pollution: A meta-analysis and systematic review." *Ecol Appl* 30 (2). doi: 10.1002/eap.2044.

Buchanan, J. B. 1971. "Pollution by synthetic fibres." *Mar Pollut Bull* 2 (2). doi: 10.1016/0025-326X(71)90136-6.

Bullard, R. 1990. *Dumping in Dixie: Race, Class, and Environmental Quality.* Boulder, CO: Westview.

Bullard, R. 2014. "The Mountains of Houston: Environmental Justice and the Politics of Garbage." Cite 93, OffCite.

Bullard, R. D., P. Mohai, R. Saha, and B. Wright. 2008. "Toxic wastes and race at twenty: Why race still matters after all of these years." *Environ Law* 38:371.

Bulleri, F., C. Ravaglioli, S. Anselmi, and M. Renzi. 2021. "The sea cucumber *Holothuria tubulosa* does not reduce the size of microplastics but enhances their resuspension in the water column." *Sci Total Environ* 781:146650. doi: 10.1016/j.scitotenv.2021.146650.

Butterworth, A., I. Clegg, and C. Bass. 2012. *Marine Debris: a Global Picture of the Impact on Animal Welfare and of Animal-Focused Solutions.* London: World Society for the Protection of Animals.

Buxton, R. T., C. A. Currey, P. O. Lyver, and C. J. Jones. 2013. "Incidence of plastic fragments among burrow-nesting seabird colonies on offshore islands in northern New Zealand." *Mar Pollut Bull* 74 (1):420-424. doi: 10.1016/j.marpolbul.2013.07.011.

Cadée, G. C. 2002. "Seabirds and floating plastic debris." *Mar Pollut Bull* 44 (11):1294-1295. doi: 10.1016/s0025-326x(02)00264-3.

California Ocean Protection Council and NOAA (National Oceanic and Atmospheric Administration) Marine Debris Program. 2018. *California Ocean Litter Prevention Strategy: Addressing Marine Debris from Source to Sea.* http://www.opc.ca.gov/webmaster/_media_library/2018/06/2018_CA_OceanLitterStrategy.pdf.

California Trash Monitoring Methods Projects. 2021. https://sites.google.com/sfei.org/trash/.

California Water Boards. 2021. "Microplastics." https://www.waterboards.ca.gov/drinking_water/certlic/drinkingwater/microplastics.html.

CalTrans. 2020. "Garbage a Growing Distraction for Caltrans." https://dot.ca.gov/programs/public-affairs/mile-marker/fall-2020/litter-removal.

Campanale, C., I. Savino, I. Pojar, C. Massarelli, and V. Felice Uricchio. 2020. "A practical overview of methodologies for sampling and analysis of microplastics in riverine environments." *Sustainability* 12 (17). doi: 10.3390/su12176755.

Campani, T., M. Baini, M. Giannetti, F. Cancelli, C. Mancusi, F. Serena, L. Marsili, S. Casini, and M. C. Fossi. 2013. "Presence of plastic debris in loggerhead turtle stranded along the Tuscany coasts of the Pelagos Sanctuary for Mediterranean Marine Mammals (Italy)." *Mar Pollut Bull* 74 (1):225-230. doi: 10.1016/j.marpolbul.2013.06.053.

Candela, A., K. Edelson, M. M. Gierach, D. R. Thompson, G. Woodward, and D. Wettergreen. 2021. "Using remote sensing and in situ measurements for efficient mapping and optimal sampling of coral reefs." *Front in Mar Sci* 8 (1195). doi: 10.3389/fmars.2021.689489.

Cannon, C. 2020. "Examining rural environmental injustice: An analysis of ruralness, class, race, and gender on the presence of landfills across the United States." *J Rural Community Dev* 15 (1).

Carlton, J. T., J. W. Chapman, J. B. Geller, J. A. Miller, D. A. Carlton, M. I. McCuller, N. C. Treneman, B. P. Steves, and G. M. Ruiz. 2017. "Tsunami-driven rafting: Transoceanic species dispersal and implications for marine biogeography." *Science* 357 (6358):1402-1406. doi: 10.1126/science.aao1498.

Carpenter, E. J., and K. L. Smith, Jr. 1972. "Plastics on the Sargasso Sea surface." *Science* 175 (4027):1240-1241. doi: 10.1126/science.175.4027.1240.

Carpenter, E. J., S. J. Anderson, G. R. Harvey, H. P. Miklas, and B. B. Peck. 1972. "Polystyrene spherules in coastal waters." *Science* 178 (4062):749-750. doi: 10.1126/science.178.4062.749.

Carr, A. 1987. "Impact of nondegradable marine debris on the ecology and survival outlook of sea turtles." *Mar Pollut Bull* 18 (6):352-356. doi: 10.1016/s0025-326x(87)80025-5.

Carr, S. A., J. Liu, and A. G. Tesoro. 2016. "Transport and fate of microplastic particles in wastewater treatment plants." *Water Res* 91:174-182. doi: 10.1016/j.watres.2016.01.002.

Carson, H. S., M. R. Lamson, D. Nakashima, D. Toloumu, J. Hafner, N. Maximenko, and K. J. McDermid. 2013. "Tracking the sources and sinks of local marine debris in Hawai'i." *Mar Environ Res* 84:76-83. doi: 10.1016/j.marenvres.2012.12.002.

Castellón, I. G. 2021. "Cancer Alley and the fight against environmental racism." *Villanova Environ Law J* 32:15.

CCME (Canadian Council of Ministers of the Environment). 2019. Canada-Wide Action Plan on Zero Plastic Waste: Phase 1. Winnipeg: CCME.

CCME. 2020. Canada-Wide Action Plan on Zero Plastic Waste: Phase 2. Winnipeg: Canadian Council of Ministers of the Environment.

Center for Biological Diversity. 2021. "Lawsuit Challenges Federal Industrial Stormwater Permit's Failure to Control U.S. Plastic Pollution, Protect Endangered Species." https://biologicaldiversity.org/w/news/press-releases/lawsuit-challenges-federal-industrial-stormwater-permits-failure-to-control-us-plastic-pollution-protect-endangered-species-2021-07-01/.

Chamas, A., H. Moon, J. Zheng, Y. Qiu, T. Tabassum, J. H. Jang, M. Abu-Omar, S. L. Scott, and S. Suh. 2020. "Degradation rates of plastics in the environment." *ACS Sustain Chem Eng* 8 (9):3494-3511. doi: 10.1021/acssuschemeng.9b06635.

Chapron, L., E. Peru, A. Engler, J. F. Ghiglione, A. L. Meistertzheim, A. M. Pruski, A. Purser, G. Vétion, P. E. Galand, and F. Lartaud. 2018. "Macro- and microplastics affect cold-water corals growth, feeding and behaviour." *Sci Rep* 8 (1). doi: 10.1038/s41598-018-33683-6.

Chellamani, K. P., D. Veerasubramanian, and R. S. Vignesh Balaji. 2013. "Surgical face masks: Manufacturing methods and classification." *J Acad Ind Res* 2 (6):320-324.

Chen, T.-L., H. Kim, S.-Y. Pan, P.-C. Tseng, Y.-P. Lin, and P.-C. Chiang. 2020. "Implementation of green chemistry principles in circular economy system towards sustainable development goals: Challenges and perspectives." *Sci Total Environ* 716:136998. doi: https://doi.org/10.1016/j.scitotenv.2020.136998.

China Government Network. 2021. "The Ministry of Ecology and Environment Answered Questions from Netizens about 'Using Special Awards and Other Methods to Strengthen the Treatment of White Plastic Waste'" [in Chinese]. http://www.gov.cn/hudong/2021-07/23/content_5626736.htm.

Choy, C. A., and J. C. Drazen. 2013. "Plastic for dinner? Observations of frequent debris ingestion by pelagic predatory fishes from the central North Pacific." *Mar Ecol Prog Ser* 485:155-163. doi: 10.3354/meps10342.

Choy, C. A., B. H. Robison, T. O. Gagne, B. Erwin, E. Firl, R. U. Halden, J. A. Hamilton, K. Katija, S. E. Lisin, C. Rolsky, and K. S. Van Houtan. 2019. "The vertical distribution and biological transport of marine microplastics across the epipelagic and mesopelagic water column." *Sci Rep* 9 (1):7843. doi: 10.1038/s41598-019-44117-2.

Chubarenko, I., and N. Stepanova. 2017. "Microplastics in sea coastal zone: Lessons learned from the Baltic amber." *Environ Pollut* 224:243-254. doi: 10.1016/j.envpol.2017.01.085.

Chubarenko, I., I. Efimova, M. Bagaeva, A. Bagaev, and I. Isachenko. 2020. "On mechanical fragmentation of single-use plastics in the sea swash zone with different types of bottom sediments: Insights from laboratory experiments." *Mar Pollut Bull* 150. doi: 10.1016/j.marpolbul.2019.110726.

Ciappa, A. C. 2021. "Marine plastic litter detection offshore Hawai'i by Sentinel-2." *Mar Pollut Bull* 168:112457. doi: 10.1016/j.marpolbul.2021.112457.

CIEL (Center for International Environmental Law). 2017. The Production of Plastic and Petrochemical Feedstocks. In *Fueling Plastics*. Washington, D.C.: Center for International Environmental Law.

CIEL. 2018. "Untested assumptions and unanswered questions in the plastics boom." In *The Long-Term Prospects for the Plastics Boom*. Washington, D.C.: Center for International Environmental Law.

CIEL. 2019. *Plastic & Climate: The Hidden Cost of a Plastic Planet*, edited by L. A. Hamilton and S. Feit. Washington, D.C.: Center for International Environmental Law.

CIEL. 2020. *Pandemic Crisis, Systemic Decline: Why Exploiting the COVID-19 Crisis Will Not Save the Oil, Gas, and Plastic Industries*. Washington, D.C.: Center for International and Environmental Law.

Closed Loop Partners. 2020. Navigating Plastic Alternatives in a Circular Economy. https://www.closedlooppartners.com/research/navigating-plastic-alternatives-in-a-circular-economy.

Coates, G. W., and Y. D. Y. L. Getzler. 2020. "Chemical recycling to monomer for an ideal, circular polymer economy." *Nat Rev Mater* 5 (7):501-516. doi: 10.1038/s41578-020-0190-4.

Coe, J. M., G. B. Antonelis, and K. Moy. 2019. "Taking control of persistent solid waste pollution." *Mar Pollut Bull* 139:105-110. doi: 10.1016/j.marpolbul.2018.12.004.

Cohen, J. H., A. M. Internicola, R. A. Mason, and T. Kukulka. 2019. "Observations and simulations of microplastic debris in a tide, wind, and freshwater-driven estuarine environment: The Delaware Bay." *Environ Sci Technol* 53 (24):14204-14211. doi: 10.1021/acs.est.9b04814.

Coish, P., E. McGovern, J. B. Zimmerman, and P. T. Anastas. 2018. "The value-adding connections between the management of ecoinnovation and the principles of green chemistry and green engineering." In *Green Chemistry*, edited by B. Török and T. Dransfield, 981-998. Amsterdam: Elsevier.

Cole, M., P. K. Lindeque, E. Fileman, J. Clark, C. Lewis, C. Halsband, and T. S. Galloway. 2016. "Microplastics alter the properties and sinking rates of zooplankton faecal pellets." *Environ Sci Technol* 50 (6):3239-3246. doi: 10.1021/acs.est.5b05905.

Coleman, C., and E. Dietz. 2019. Fact Sheet | "Fossil Fuel Subsidies: A Closer Look at Tax Breaks and Societal Costs. White Paper. Environmental and Energy Study Institute.

Comtrade, UN. 2020. "All Commodities." https://comtrade.un.org/db/mr/daCommodities Results.aspx?px=h2&cc=TOTAL.

Connors, K. A., S. D. Dyer, and S. E. Belanger. 2017. "Advancing the quality of environmental microplastic research." *Environ Toxicol Chem* 36 (7):1697-1703. doi: 10.1002/etc.3829.

Cornago, E., P. Börkey, and A. Brown. 2021. "Preventing Single-use Plastic Waste." OECD Environment Working Paper No. 182. Paris: OECD Publishing.

CounterMEASURE. 2021. "CountMEASURE for Plastic Free Rivers." https://countermeasure. asia/.

Cowger, W., A. M. Booth, B. M. Hamilton, C. Thaysen, S. Primpke, K. Munno, A. L. Lusher, A. Dehaut, V. P. Vaz, and M. Liboiron. 2020. "Reporting guidelines to increase the reproducibility and comparability of research on microplastics." *Appl Spectrosc* 74 (9):1066-1077.

Cózar, A., F. Echevarría, J. I. González-Gordillo, X. Irigoien, B. Úbeda, S. Hernández-León, A. T. Palma, S. Navarro, J. García-de-Lomas, A. Ruiz, M. L. Fernández-de-Puelles, and C. M. Duarte. 2014. "Plastic debris in the open ocean." *Proc Natl Acad Sci USA* 111 (28):10239-10244. doi: 10.1073/pnas.1314705111.

Curren, E., and S. C. Y. Leong. 2019. "Profiles of bacterial assemblages from microplastics of tropical coastal environments." *Sci Total Environ* 655:313-320. doi: 10.1016/j.scitotenv.2018.11.250.

Daly, H. 2019. "Growthism: Its ecological, economic and ethical limits." *Real-World Econ Rev* (87):9-22.

Dameron, O. J., M. Parke, M. A. Albins, and R. Brainard. 2007. "Marine debris accumulation in the Northwestern Hawaiian Islands: An examination of rates and processes." *Mar Pollut Bull* 54 (4):423-433. doi: 10.1016/j.marpolbul.2006.11.019.

Damgacioglu, H., M. Hornilla, O. Bafail, and N. Celik. 2020. "Recovering value from single stream material recovery facilities - An outbound contamination analysis in Florida." *Waste Manag* 102:804-814. doi: 10.1016/j.wasman.2019.11.020.

Davies, T. 2018. "Toxic space and time: Slow violence, necropolitics, and petrochemical pollution." *Ann Am Assoc Geogr* 108 (6):1537-1553. doi: 10.1080/24694452.2018.1470924.

Dawson, A. L., S. Kawaguchi, C. K. King, K. A. Townsend, R. King, W. M. Huston, and S. M. Bengtson Nash. 2018. "Turning microplastics into nanoplastics through digestive fragmentation by Antarctic krill." *Nat Commun* 9 (1):1001. doi: 10.1038/s41467-018-03465-9.

Dawson, A. L., M. F. M. Santana, M. E. Miller, and F. J. Kroon. 2021. "Relevance and reliability of evidence for microplastic contamination in seafood: A critical review using Australian consumption patterns as a case study." *Environ Pollut* 276. doi: 10.1016/j.envpol.2021.116684.

De Blasio, N., and P. Fallon. 2022. "Avoiding a plastic pandemic: The future of sustainability in a post COVID-19 world." *J Self-Gov Manag Econ* 10 (1):7-29. doi: 10.22381/jsme10120221.

de la Torre, G. E., D. C. Dioses-Salinas, C. I. Pizarro-Ortega, and L. Santillán. 2021. "New plastic formations in the Anthropocene." *Sci Total Environ* 754:142216. doi: 10.1016/j.scitotenv.2020.142216.

de Stephanis, R., J. Giménez, E. Carpinelli, C. Gutierrez-Exposito, and A. Cañadas. 2013. "As main meal for sperm whales: Plastics debris." *Mar Pollut Bull* 69 (1-2):206-214. doi: 10.1016/j.marpolbul.2013.01.033.

Deanin, R. D. 1975. "Additives in plastics." *Environ Health Perspect* 11:35-39.

Deidun, A., A. Gauci, S. Lagorio, and F. Galgani. 2018. "Optimising beached litter monitoring protocols through aerial imagery." *Mar Pollut Bull* 131 (Pt A):212-217. doi: 10.1016/j.marpolbul.2018.04.033.

Dell'Ariccia, G., R. A. Phillips, J. A. van Franeker, N. Gaidet, P. Catry, J. P. Granadeiro, P. G. Ryan, and F. Bonadonna. 2017. "Comment on 'Marine plastic debris emits a keystone infochemical for olfactory foraging seabirds' by Savoca et al." *Sci Adv* 3 (6):e1700526. doi: 10.1126/sciadv.1700526.

Desforges, J. P., M. Galbraith, and P. S. Ross. 2015. "Ingestion of microplastics by zooplankton in the Northeast Pacific Ocean." *Arch Environ Contam Toxicol* 69 (3):320-330. doi: 10.1007/s00244-015-0172-5.

Dharmaraj, S., V. Ashokkumar, S. Hariharan, A. Manibharathi, P. L. Show, C. T. Chong, and C. Ngamcharussrivichai. 2021. "The COVID-19 pandemic face mask waste: A blooming threat to the marine environment." *Chemosphere* 272:129601. doi: 10.1016/j.chemosphere.2021.129601.

Di Beneditto, A. P., and R. M. Ramos. 2014. "Marine debris ingestion by coastal dolphins: What drives differences between sympatric species?" *Mar Pollut Bull* 83 (1):298-301. doi: 10.1016/j.marpolbul.2014.03.057.

Donohue, M. J. 2005. "Eastern Pacific Ocean source of Northwestern Hawaiian Islands marine debris supported by errant fish aggregating device." *Mar Pollut Bull* 50 (8):886-888. doi: 10.1016/j.marpolbul.2005.05.003.

Donohue, M. J., and D. G. Foley. 2007. "Remote sensing reveals links among the endangered Hawaiian monk seal, marine debris, and El Niño." *Mar Mamm Sci* 23 (2):468-473. doi: 10.1111/j.1748-7692.2007.00114.x.

Donohue, M. J., R. C. Boland, C. M. Sramek, and G. A. Antonelis. 2001. "Derelict fishing gear in the Northwestern Hawaiian Islands: Diving surveys and debris removal in 1999 confirm threat to coral reef ecosystems." *Mar Pollut Bull* 42 (12):1301-1312. doi: 10.1016/s0025-326x(01)00139-4.

Donohue, M. J., J. Masura, T. Gelatt, R. Ream, J. D. Baker, K. Faulhaber, and D. T. Lerner. 2019. "Evaluating exposure of northern fur seals, *Callorhinus ursinus*, to microplastic pollution through fecal analysis." *Mar Pollut Bull* 138:213-221. doi: 10.1016/j.marpolbul.2018.11.036.

Doyle, M. J., W. Watson, N. M. Bowlin, and S. B. Sheavly. 2011. "Plastic particles in coastal pelagic ecosystems of the Northeast Pacific ocean." *Mar Environ Res* 71 (1):41-52. doi: 10.1016/j.marenvres.2010.10.001.

Dris, R., J. Gasperi, and B. Tassin. 2018. "Sources and fate of microplastics in urban areas: A focus on Paris megacity." In *Freshwater Microplastics: Emerging Environmental Contaminants?* edited by M. Wagner and S. Lambert. New York: Springer International. doi: 10.1007/978-3-319-61615-5_4.

Dzhanova, Y. 2020. "Sanitation workers battle higher waste levels in residential areas as coronavirus outbreak persists." *CNBC*. https://www.cnbc.com/2020/05/16/coronavirus-sanitation-workers-battle-higher-waste-levels.html.

Earn, A., K. Bucci, and C. M. Rochman. 2021. "A systematic review of the literature on plastic pollution in the Laurentian Great Lakes and its effects on freshwater biota." *J Great Lakes Res* 47 (1):120-133. doi: 10.1016/j.jglr.2020.11.001.

Earp, H. S., and A. Liconti. 2020. "Science for the future: The use of citizen science in marine research and conservation." In *YOUMARES 9 - The Oceans: Our Research, Our Future*, edited by S. Jungblut, V. Liebich, and M. Bode-Dalby, 1-19. New York: Springer International. doi: 10.1007/978-3-030-20389-4_1.

Ebbesmeyer, C. C., W. J. Ingraham, J. A. Jones, and M. J. Donohue. 2012. "Marine debris from the Oregon Dungeness crab fishery recovered in the Northwestern Hawaiian Islands: Identification and oceanic drift paths." *Mar Pollut Bull* 65 (1-3):69-75. doi: 10.1016/j.marpolbul.2011.09.037.

Edwards, E. 2020. "How Are Goggles Made?" Thomas Publishing Company. https://www.thomasnet.com/articles/plant-facility-equipment/how-are-goggles-made/.

Efimova, I., M. Bagaeva, A. Bagaev, A. Kileso, and I. P. Chubarenko. 2018. "Secondary microplastics generation in the sea swash zone with coarse bottom sediments: Laboratory experiments." *Front Mar Sci* 5. doi: 10.3389/fmars.2018.00313.

Elert, A. M., R. Becker, E. Duemichen, P. Eisentraut, J. Falkenhagen, H. Sturm, and U. Braun. 2017. "Comparison of different methods for MP detection: What can we learn from them, and why asking the right question before measurements matters?" *Environ Pollut* 231 (Pt 2):1256-1264. doi: 10.1016/j.envpol.2017.08.074.

Ellen MacArthur Foundation. 2017. "Plastics Pact Network: National and Regional Initiatives Working Towards a Circular Economy for Plastic." https://www.ellenmacarthurfoundation.org/our-work/activities/new-plastics-economy/plastics-pact.

Enders, K., R. Lenz, C. A. Stedmon, and T. G. Nielsen. 2015. "Abundance, size and polymer composition of marine microplastics ≥10 μm in the Atlantic Ocean and their modelled vertical distribution." *Mar Pollut Bull* 100 (1):70-81.

Energy.gov. 2020. "U.S. Department of Energy Announces $27 Million in Plastics Recycling Research and Development." October 15. https://www.energy.gov/articles/us-department-energy-announces-27-million-plastics-recycling-research-and-development.

Engler, R. E. 2012. "The complex interaction between marine debris and toxic chemicals in the ocean." *Environ Sci Technol* 46 (22):12302-12315. doi: 10.1021/es3027105.

Environment and Climate Change Canada and Health Canada. 2020. Science Assessment of Plastic Pollution. Gatineau, Quebec: Environment and Climate Change Canada, Health Canada.

EOA, Inc. 2014. *San Francisco Bay Area Trash Generation Rates, Final Technical Report.* Bay Area Stormwater Management Agencies Association.

EREF (Environmental Research & Education Foundation). 2016. Municipal Solid Waste Management in the U.S.: 2010 & 2013.

Eriksen, M., L. C. Lebreton, H. S. Carson, M. Thiel, C. J. Moore, J. C. Borerro, F. Galgani, P. G. Ryan, and J. Reisser. 2014. "Plastic pollution in the world's oceans: More than 5 trillion plastic pieces weighing over 250,000 tons afloat at sea." *PLoS ONE* 9 (12):e111913. doi: 10.1371/journal.pone.0111913.

Eriksson, C., and H. Burton. 2003. "Origins and biological accumulation of small plastic particles in fur seals from Macquarie Island." *Ambio* 32 (6):380-384. doi: 10.1579/0044-7447-32.6.380.

Erogunaiye, O. 2019. "Environmental Justice and Landfills: Application of Spatial and Statistical Analysis for Assessing Landfill Sites in Texas." Texas Southern University.

European Bioplastics. 2020. "Bioplastics Market Development Update 2020." European Bioplastics Conference. https://docs.european-bioplastics.org/conference/Report_Bioplastics_Market_Data_2020_short_version.pdf.

European Commission. 2018. "Annexes to the Communication from the Commission to the European Parliament, the Council, the European Economic and Social Committee and the Committee of the Regions: A European Strategy for Plastics in a Circular Economy."

European Commission. 2020. "Circular Economy Action Plan: The EU's New Circular Action Plan Paves the Way for a Cleaner and More Competitive Europe." https://ec.europa.eu/environment/strategy/circular-economy-action-plan_en.

European Commission. 2021. "Circular Economy Action Plan." https://ec.europa.eu/environment/strategy/circular-economy-action-plan_en.

European Environment Agency. 2021. *Plastics, the Circular Economy and Europe's Environment: A Priority for Action.* Luxembourg: European Union.

Evans, M. C., and C. S. Ruf. 2021. "Toward the detection and imaging of ocean microplastics with a spaceborne radar." *IEEE Trans Geosci Remote Sens*: 1-9. doi: 10.1109/tgrs.2021.3081691.

Fallati, L., A. Polidori, C. Salvatore, L. Saponari, A. Savini, and P. Galli. 2019. "Anthropogenic marine debris assessment with unmanned aerial vehicle imagery and deep learning: A case study along the beaches of the Republic of Maldives." *Sci Total Environ* 693:133581. doi: 10.1016/j.scitotenv.2019.133581.

Farady, S. E. 2019. "Microplastics as a new, ubiquitous pollutant: Strategies to anticipate management and advise seafood consumers." *Mar Policy* 104:103-107. doi: 10.1016/j.marpol.2019.02.020.

Fernández-Llamazares, Á., M. Garteizgogeascoa, N. Basu, E. S. Brondizio, M. Cabeza, J. Martínez-Alier, P. McElwee, and V. Reyes-García. 2020. "A state-of-the-art review of Indigenous peoples and environmental pollution." *Integr Environ Assess Manag* 16 (3):324-341. doi: 10.1002/ieam.4239.

Ford, H. V., N. H. Jones, A. J. Davies, B. J. Godley, J. R. Jambeck, I. E. Napper, C. C. Suckling, G. J. Williams, L. C. Woodall, and H. J. Koldewey. 2022. "The fundamental links between climate change and marine plastic pollution." *Sci Total Environ* 806:150392. doi: https://doi.org/10.1016/j.scitotenv.2021.150392.

Franklin Associates. 2014. *Impact of Plastics Packaging on Life Cycle Energy Consumption and Greenhouse Gas Emissions in the United States and Canada.* Overland Park, Franklin Associates, A Division of ERG.

Fujieda, S. 2007. "Marine litter on the sea bottom in Hakata Bay." *Mem Fac Fish Kagoshima Univ* 56:67-74.

G20. 2021. "The United States: Actions and Progress on Marine Plastic Litter." https://g20mpl.org/partners/unitedstates.

Gago, J., F. Galgani, T. Maes, and R. C. Thompson. 2016. "Microplastics in seawater: Recommendations from the Marine Strategy Framework Directive Implementation Process." *Front Mar Sci* 3. doi: 10.3389/fmars.2016.00219.

GAIA (Global Alliance for Incinerator Alternatives). 2019. "Discarded: Communities on the Frontlines of the Global Plastic Crisis." Berkeley, CA: GAIA.

Galgani, F., A. Souplet, and Y. Cadiou. 1996. "Accumulation of debris on the deep sea floor off the French Mediterranean coast." *Mar Ecol Prog Ser* 142:225-234. doi: 10.3354/meps142225.

Galgani, L., and S. A. Loiselle. 2021. "Plastic pollution impacts on marine carbon biogeochemistry." *Environ Pollut* 268:115598. doi: https://doi.org/10.1016/j.envpol.2020.115598.

Gall, S. C., and R. C. Thompson. 2015. "The impact of debris on marine life." *Mar Pollut Bull* 92 (1-2):170-179. doi: 10.1016/j.marpolbul.2014.12.041.

Garaba, S. P., and H. M. Dierssen. 2018. "An airborne remote sensing case study of synthetic hydrocarbon detection using short wave infrared absorption features identified from marine-harvested macro- and microplastics." *Remote Sens Environ* 205:224-235.

Garaba, S. P., T. Acuña-Ruz, and C. B. Mattar. 2020. "Hyperspectral longwave infrared reflectance spectra of naturally dried algae, anthropogenic plastics, sands and shells." *Earth Sys Sci Data* 12 (4):2665-2678.

Gasperi, J., R. Dris, T. Bonin, V. Rocher, and B. Tassin. 2014. "Assessment of floating plastic debris in surface water along the Seine River." *Environ Pollut* 195:163-166. doi: 10.1016/j.envpol.2014.09.001.

GDOT (Georgia Department of Transportation). 2020. "Keep It Clean, Georgia: Fact Sheets." https://keepitcleanga-gdot.hub.arcgis.com/pages/print.

Ge, Z., H. Shi, X. Mei, Z. Dai, and D. Li. 2016. "Semi-automatic recognition of marine debris on beaches." *Sci Rep* 6:25759. doi: 10.1038/srep25759.

Geissdoerfer, M., P. Savaget, N. M. P. Bocken, and E. Jan Hultink. 2017. "The Circular Economy – A new sustainability paradigm?" *J Clean Prod* 143:757-768. doi: 10.1016/j.jclepro.2016.12.048.

Gentry, R. L., and G. L. Kooyman. 1986. *Fur Seals: Maternal Strategies on Land and at Sea.* Vol. 64. Princeton University Press.

GESAMP (Joint Group of Experts on the Scientific Aspects of Marine Environmental Protection). 2019. *Guidelines or the Monitoring and Assessment of Plastic Litter and Microplastics in the Ocean,* edited by P. J. Kershaw, A. Turra, and F. Galgani.

Geyer, R., J. R. Jambeck, and K. L. Law. 2017. "Production, use, and fate of all plastics ever made." *Sci Adv* 3 (7):e1700782. doi: 10.1126/sciadv.1700782.

Ghosal, S., M. Chen, J. Wagner, Z. M. Wang, and S. Wall. 2018. "Molecular identification of polymers and anthropogenic particles extracted from oceanic water and fish stomach – A Raman micro-spectroscopy study." *Environ Pollut* 233:1113-1124. doi: 10.1016/j.envpol.2017.10.014.

Goddijn-Murphy, L., and B. Williamson. 2019. "On thermal infrared remote sensing of plastic pollution in natural waters." *Remote Sens* 11 (18):2159.

Goldberg, E. D. 1997. "Plasticizing the seafloor: An overview." *Environ Technol* 18 (2):195-201. doi: 10.1080/09593331808616527.

Goldstein, M. C., A. J. Titmus, and M. Ford. 2013. "Scales of spatial heterogeneity of plastic marine debris in the Northeast Pacific Ocean." *PLoS ONE* 8 (11):e80020. doi: 10.1371/journal.pone.0080020.

Goncalves, G., U. Andriolo, L. Pinto, and F. Bessa. 2020. "Mapping marine litter using UAS on a beach-dune system: A multidisciplinary approach." *Sci Total Environ* 706:135742. doi: 10.1016/j.scitotenv.2019.135742.

González-Fernández, D., and G. Hanke. 2017. "Toward a harmonized approach for monitoring of riverine floating macro litter inputs to the marine environment." *Front Mar Sci* 4. doi: 10.3389/fmars.2017.00086.

González-Fernández, D., A. Cózar, G. Hanke, J. Viejo, C. Morales-Caselles, R. Bakiu, D. Barceló, F. Bessa, A. Bruge, M. Cabrera, J. Castro-Jiménez, M. Constant, R. Crosti, Y. Galletti, A. E. Kideys, N. Machitadze, J. Pereira de Brito, M. Pogojeva, N. Ratola, J. Rigueira, E. Rojo-Nieto, O. Savenko, R. I. Schöneich-Argent, G. Siedlewicz, G. Suaria, and M. Tourgeli. 2021. "Floating macrolitter leaked from Europe into the ocean." *Nat Sustain* 4 (6):474-483. doi: 10.1038/s41893-021-00722-6.

Goulder, L. H., and D. Kennedy. 1997. *Valuing Ecosystem Services: Philosophical Bases and Empirical Methods*. Washington, D.C.: Island Press.

Gove, J. M., J. L. Whitney, M. A. McManus, J. Lecky, F. C. Carvalho, J. M. Lynch, J. Li, P. Neubauer, K. A. Smith, J. E. Phipps, D. R. Kobayashi, K. B. Balagso, E. A. Contreras, M. E. Manuel, M. A. Merrifield, J. J. Polovina, G. P. Asner, J. A. Maynard, and G. J. Williams. 2019. "Prey-size plastics are invading larval fish nurseries." *Proc Nat Acad Sci USA* 116 (48):24143-24149. doi: 10.1073/pnas.1907496116.

Government of Canada. 2021. "Zero Plastic Waste: Canada's Actions." https://www.canada.ca/en/environment-climate-change/services/managing-reducing-waste/reduce-plastic-waste/canada-action.html.

Gray, A. D., H. Wertz, R. R. Leads, and J. E. Weinstein. 2018. "Microplastic in two South Carolina estuaries: Occurrence, distribution, and composition." *Mar Pollut Bull* 128:223-233. doi: 10.1016/j.marpolbul.2018.01.030.

GreenFacts. 2021. "Non-market Value." https://www.greenfacts.org/glossary/mno/non-market-value.htm.

Greenpeace. 2020. *Circular Claims Fall Flat: Comprehensive U.S. Survey of Plastics Recyclability*, edited by J. Hocevar, I. Schlegel, and P. Wheeler. Washington, D.C.: Greenpeace.

Gregory, M. R. 2009. "Environmental implications of plastic debris in marine settings—entanglement, ingestion, smothering, hangers-on, hitch-hiking and alien invasions." *Philos Trans R Soc Lond B Biol Sci* 364 (1526):2013-2025. doi: 10.1098/rstb.2008.0265.

Guangdong Provincial Development and Reform Commission. 2020. "Picture Reading | Interpretation of the policy of 'Implementation Opinions on Further Strengthening the Treatment of Plastic Pollution.'" http://drc.gd.gov.cn/zcjd5635/content/post_3073286.html.

Guven, O., K. Gökdağ, and A. Kideys. 2016. "Microplastic densities in seawater and sediment from the north eastern Mediterranean Sea." Paper presented at Türkiye Deniz Bilimleri Konferansı, Ankara, Turkey.

Guzzetti, E., A. Sureda, S. Tejada, and C. Faggio. 2018. "Microplastic in marine organism: Environmental and toxicological effects." *Environ Toxicol Pharmacol* 64:164-171. doi: 10.1016/j.etap.2018.10.009.

Habel, S., C. H. Fletcher, K. Rotzoll, and A. I. El-Kadi. 2017. "Development of a model to simulate groundwater inundation induced by sea-level rise and high tides in Honolulu, Hawaii." *Water Res* 114:122-134. doi: 10.1016/j.watres.2017.02.035.

Haberstroh, C. J., M. E. Arias, Z. Yin, and M. C. Wang. 2021. "Effects of hydrodynamics on the cross-sectional distribution and transport of plastic in an urban coastal river." *Water Environ Res* 93 (2):186-200. doi: 10.1002/wer.1386.

Hagen, P., R. LaMotte, and D. Meng. 2021. "The circular economy runs through Basel." *Environ Forum.* https://www.bdlaw.com/content/uploads/2021/08/SeptOct2021 _The-Environmental-Forum_LeadFeature.pdf.

Halden, R. U. 2010. "Plastics and health risks." *Annu Rev Public Health* 31:179-194. doi: 10.1146/annurev.publhealth.012809.103714.

Hamilton, B. M., C. M. Rochman, T. J. Hoellein, B. H. Robison, K. S. Van Houtan, and C. A. Choy. 2021. "Prevalence of microplastics and anthropogenic debris within a deep-sea food web." *Mar Ecol Prog Ser* 675:23-33. doi: 10.3354/meps13846.

Hamilton, S., and G. B. Baker. 2019. "Technical mitigation to reduce marine mammal bycatch and entanglement in commercial fishing gear: Lessons learnt and future directions." *Rev Fish Biol Fish* 29 (2):223-247. doi: 10.1007/s11160-019-09550-6.

Hanvey, J. S., P. J. Lewis, J. L. Lavers, N. D. Crosbie, K. Pozo, and B. O. Clarke. 2017. "A review of analytical techniques for quantifying microplastics in sediments." *Anal Meth* 9 (9):1369-1383. doi: 10.1039/c6ay02707e.

Hardesty, B. D., C. Wilcox, Q. Schuyler, T. J. Lawson, and K. Opie. 2017. *Developing a Baseline Estimate of Amounts, Types, Sources and Distribution of Coastal Litter—An Analysis of US Marine Debris Data.* Version 1.2. CSIRO: EP167399. https://marinedebris.noaa.gov/ sites/default/files/publications-files/An_analysis_of_marine_debris_in_the_US _FINAL_REP.pdf.

Harrison, J. P., M. Sapp, M. Schratzberger, and A. M. Osborn. 2011. "Interactions between microorganisms and marine micro plastics: A call for research." *Mar Technol Soc J* 45 (2):12-20. doi: 10.4031/MTSJ.45.2.2.

Hartmann, N. B., T. Hüffer, R. C. Thompson, M. Hassellöv, A. Verschoor, A. E. Daugaard, S. Rist, T. Karlsson, N. Brennholt, M. Cole, M. P. Herrling, M. C. Hess, N. P. Ivleva, A. L. Lusher, and M. Wagner. 2019. "Are we speaking the same language? Recommendations for a definition and categorization framework for plastic debris." *Environ Sci Technol* 53 (3):1039-1047. doi: 10.1021/acs.est.8b05297.

He, P., and P. Suuronen. 2018. "Technologies for the marking of fishing gear to identify gear components entangled on marine animals and to reduce abandoned, lost or otherwise discarded fishing gear." *Mar Pollut Bull* 129 (1):253-261. doi: 10.1016/j.marpolbul.2018.02.033.

Heller, M. C., M. H. Mazor, and G. A. Keoleian. 2020. "Plastics in the US: Toward a material flow characterization of production, markets and end of life." *Environ Res Lett* 15 (9). doi: 10.1088/1748-9326/ab9e1e.

Henderson, J. R. 2001. "A pre- and post-MARPOL Annex V summary of Hawaiian monk seal entanglements and marine debris accumulation in the northwestern Hawaiian Islands, 1982–1998." *Mar Pollut Bull* 42 (7):584-589. doi: 10.1016/s0025-326x(00)00204-6.

Henneberry, B. 2020. "How Surgical Masks Are Made." Thomas Publishing Company.

Hermsen, E., S. M. Mintenig, E. Besseling, and A. A. Koelmans. 2018. "Quality criteria for the analysis of microplastic in biota samples: A critical review." *Environ Sci Technol* 52 (18):10230-10240. doi: 10.1021/acs.est.8b01611.

Hess, N. A., C. A. Ribic, and I. Vining. 1999. "Benthic marine debris, with an emphasis on fishery-related items, surrounding Kodiak Island, Alaska, 1994–1996." *Mar Pollut Bull* 38 (10):885-890. doi: 10.1016/s0025-326x(99)00087-9.

Hidalgo-Ruz, V., L. Gutow, R. C. Thompson, and M. Thiel. 2012. "Microplastics in the marine environment: A review of the methods used for identification and quantification." *Environ Sci Technol* 46 (6):3060-75. doi: 10.1021/es2031505.

HM Government. 2018. *A Green Future: Our 25 Year Plan to Improve the Environment.* London: Crown. https://www.gov.uk/government/publications/25-year-environment-plan.

Hockenos, P. 2021. "Bold single-use plstic ban kicks Europe's plastic purge into high gear." *YaleEnvironment 360.* https://e360.yale.edu/features/europes-drive-to-slash-plastic-waste-moves-into-high-gear.

Hoellein, T. J., and C. M. Rochman. 2021. "The 'plastic cycle': A watershed-scale model of plastic pools and fluxes." *Front Ecol Environ* 19 (3):176-183. doi: 10.1002/fee.2294.

Hoellein, T. J., A. R. McCormick, J. Hittie, M. G. London, J. W. Scott, and J. J. Kelly. 2017. "Longitudinal patterns of microplastic concentration and bacterial assemblages in surface and benthic habitats of an urban river." *Freshwater Sci* 36 (3):491-507. doi: 10.1086/693012.

Hoffman, M. J., and E. Hittinger. 2017. "Inventory and transport of plastic debris in the Laurentian Great Lakes." *Mar Pollut Bull* 115 (1-2):273-281. doi: 10.1016/j.marpolbul.2016.11.061.

Hoornweg, D., and P. Bhada-Tata. 2012. *What A Waste: A Global Review of Solid Waste Management*. Washington, DC: World Bank.

Hopewell, J., R. Dvorak, and E. Kosior. 2009. "Plastics recycling: Challenges and opportunities." *Philos Trans Soc Biol Sci* 364 (1526):2115-2126.

Howell, E. A., S. J. Bograd, C. Morishige, M. P. Seki, and J. J. Polovina. 2012. "On North Pacific circulation and associated marine debris concentration." *Mar Pollut Bull* 65 (1-3):16-22. doi: 10.1016/j.marpolbul.2011.04.034. https://eur-lex.europa.eu/resource.html?uri=cellar:2df5d1d2-fac7-11e7-b8f5-01aa75ed71a1.0001.02/DOC_2&format=PDF.

Hu, C. 2021. "Remote detection of marine debris using satellite observations in the visible and near infrared spectral range: Challenges and potentials." *Remote Sens Environ* 259. doi: 10.1016/j.rse.2021.112414.

Hu, C., L. Feng, R. F. Hardy, and E. J. Hochberg. 2015. "Spectral and spatial requirements of remote measurements of pelagic Sargassum macroalgae." *Remote Sens Environ* 167: 229-246. doi: 10.1016/j.rse.2015.05.022.

Hung, C., N. Klasios, X. Zhu, M. Sedlak, R. Sutton, and C. M. Rochman. 2021. "Methods matter: Methods for sampling microplastic and other anthropogenic particles and their implications for monitoring and ecological risk assessment." *Integr Environ Assess Manag* 17 (1):282-291. doi: 10.1002/ieam.4325.

Hurley, R., J. Woodward, and J. J. Rothwell. 2018. "Microplastic contamination of river beds significantly reduced by catchment-wide flooding." *Nat Geosci* 11 (4):251-257. doi: 10.1038/s41561-018-0080-1.

IEA (International Energy Agency). 2018. The Future of Petrochemicals: Towards a More Sustainable Chemical Industry. https://www.iea.org/reports/the-future-of-petrochemicals.

IEA Bioenergy. 2018. *Standards and Labels Related to Biobased Products: Developments in the 2016-2018 Triennium*. Paris, IEA Bioenergy. https://www.ieabioenergy.com/blog/publications/standards-and-labels-related-to-biobased-products-developments-in-the-2016-2018-triennium.

Im, J., S. Joo, Y. Lee, B. Y. Kim, and T. Kim. 2020. "First record of plastic debris ingestion by a fin whale (*Balaenoptera physalus*) in the sea off East Asia." *Mar Pollut Bull* 159:111514. doi: 10.1016/j.marpolbul.2020.111514.

IMF (International Monetary Fund). 2019. *Global Fossil Fuel Subsidies Remain Large: An Update Based on Country-Level Estimates*. Working Paper No. 19/89, edited by D. Coady, I. Parry, N. P. Le and B. Shang.

Ingraham W. J., Jr., and C. C. Ebbesmeyer. 2001. "Surface current concentration of floating marine debris in the North Pacific Ocean: 12-year OSCURS model experiments." In *Proceedings of the International Conference on Derelict Fishing Gear and the Ocean Environment*, edited by N. McIntosh, K. Simonds, M. Donohue, C. Brammer, S. Mason, and S. Carbajal, 91-115. National Marine Sanctuaries and National Oceanic and Atmospheric Administration..

International Coastal Cleanup. 2018. *Building a Clean Swell: 2018 Report*. Washington, DC: Ocean Conservancy.

International Coastal Cleanup. 2019. *The Beach and Beyond: 2019 Report*. Washington, DC: Ocean Conservancy.

International Coastal Cleanup. 2020. *Together, We Are Team Ocean: 2020 Report*. Washington, DC: Ocean Conservancy.

International Maritime Organization. 2019. *Hull Scrapings and Marine Coatings as a Source of Microplastics*.

INTERPOL. 2020. *Strategic Analysis Report: Emerging Criminal Trends in the Global Plastic Waste Market Since January 2018*. Lyon, France: INTERPOL.

Ioakeimidis, C., C. Zeri, H. Kaberi, M. Galatchi, K. Antoniadis, N. Streftaris, F. Galgani, E. Papathanassiou, and G. Papatheodorou. 2014. "A comparative study of marine litter on the seafloor of coastal areas in the eastern Mediterranean and Black Seas." *Mar Pollut Bull* 89 (1-2):296-304. doi: 10.1016/j.marpolbul.2014.09.044.

IRP (International Resource Panel). 2021. Policy Options to Eliminate Additional Marine Plastic Litter by 2050 Under the G20 Osaka Blue Ocean Vision, edited by S. Fletcher, K. P. Roberts, Y. Shiran, J. Virdin, C. Brown, E. Buzzi, I. C. Alcolea, L. Henderson, F. Laubinger, L. Milà i Canals, S. Salam, S. A. Schmuck, J. M. Veiga, S. Winton, and K. M. Youngblood. Nairobi, Kenya: United Nations Environment Programme.

Issifu, I., E. W. Deffor, and U. R. Sumaila. 2021. "How COVID-19 could change the economics of the plastic recycling sector." *Recycling* 6 (4). doi: 10.3390/recycling6040064.

Ivar do Sul, J. A., M. F. Costa, J. S. Silva-Cavalcanti, and M. C. Araújo. 2014. "Plastic debris retention and exportation by a mangrove forest patch." *Mar Pollut Bull* 78 (1-2):252-257. doi: 10.1016/j.marpolbul.2013.11.011.

Jacobsen, J. K., L. Massey, and F. Gulland. 2010. "Fatal ingestion of floating net debris by two sperm whales (*Physeter macrocephalus*)." *Mar Pollut Bull* 60 (5):765-767. doi: 10.1016/j.marpolbul.2010.03.008.

Jambeck, J. R., and K. Johnsen. 2015. "Citizen-based litter and marine debris data collection and mapping." *Comput Sci Eng* 17 (4):20-26. doi: 10.1109/mcse.2015.67.

Jambeck, J. R., R. Geyer, C. Wilcox, T. R. Siegler, M. Perryman, A. Andrady, R. Narayan, and K. L. Law. 2015. "Plastic waste inputs from land into the ocean." *Science* 347 (6223):768-771. doi: 10.1126/science.1260352.

Jambeck, J., B. D. Hardesty, A. L. Brooks, T. Friend, K. Teleki, J. Fabres, Y. Beaudoin, A. Bamba, J. Francis, A. J. Ribbink, T. Baleta, H. Bouwman, J. Knox, and C. Wilcox. 2018. "Challenges and emerging solutions to the land-based plastic waste issue in Africa." *Mar Policy* 96:256-263. doi: 10.1016/j.marpol.2017.10.041.

Jambeck, J., E. Moss, B. Dubey, Z. Arifin, L. Godfrey, B. D. Hardesty, I. G. Hendrawan, T. T. Hien, L. Junguo, M. Matlock, S. Pahl, K. Raubenheimer, M. Thiel, R. Thompson, and L. Woodall. 2020. *Leveraging Multi-Target Strategies to Address Plastic Pollution in the Context of an Already Stressed Ocean*. Washington, DC: World Resources Institute.

Jarosova, A., J. Harazim, P. Suchy, L. Kratka, and V. Stancova. 2009. "The distribution and accumulation of phthalates in the organs and tissues of chicks after the administration of feedstuffs with different phthalate concentrations." *Vet Med (Praha)* 54 (9):427-434. doi: 10.17221/2/2009-VETMED.

Johnson, A., G. Salvador, J. Kenney, J. Robbins, S. Kraus, S. Landry, and P. Clapham. 2005. "Fishing gear involved in entanglements of right and humpback whales." *Mar Mamm Sci* 21 (4):635-645. doi: 10.1111/j.1748-7692.2005.tb01256.x.

Kanhai, L. D. K., C. Johansson, J. P. G. L. Frias, K. Gardfeldt, R. C. Thompson, and I. O'Connor. 2019. "Deep sea sediments of the Arctic Central Basin: A potential sink for microplastics." *Deep-Sea Res I: Oceanogr Res Pap* 145:137-142. doi: 10.1016/j.dsr.2019.03.003.

Kappler, A., D. Fischer, S. Oberbeckmann, G. Schernewski, M. Labrenz, K. J. Eichhorn, and B. Voit. 2016. "Analysis of environmental microplastics by vibrational microspectroscopy: FTIR, Raman or both?" *Anal Bioanal Chem* 408 (29):8377-8391. doi: 10.1007/s00216-016-9956-3.

Karasik, R., T. Vegh, Z. Diana, J. Bering, J. Caldas, A. Pickle, D. Rittschof, and J. Virdin. 2020. *20 Years of Government Responses to the Global Plastic Pollution Problem: The Plastics Policy Inventory.* Durham, NC: Nicholas Institute for Environmental Policy Solutions, Duke University.

Kartar, S., R. A. Milne, and M. Sainsbury. 1973. "Polystyrene waste in the Severn Estuary." *Mar Pollut Bull* 4 (9). doi: 10.1016/0025-326x(73)90010-6.

Katija, K., C. A. Choy, R. E. Sherlock, A. D. Sherman, and B. H. Robison. 2017. "From the surface to the seafloor: How giant larvaceans transport microplastics into the deep sea." *Sci Adv* 3 (8):e1700715. doi: 10.1126/sciadv.1700715.

Kaza, S., L. Yao, P. Bhada-Tata, and F Van Woerden. 2018. *What a Waste 2.0: A Global Snapshot of Solid Waste Management to 2050.* Washington, D.C.: World Bank.

Keep America Beautiful Inc. 2010. "Litter in America: Results from the Nation's Largest Litter Study." https://kab.org/wp-content/uploads/2019/11/LitterinAmerica_FactSheet _LitterOverview.pdf.

Keller & Heckman. 2020. "CA to Require Minimum Recycled Content in Plastic Bottles." packaginglaw.com. https://www.packaginglaw.com/news/ca-require-minimum-recycled-content-plastic-bottles.

Kemp-Benedict, E., and S. Kartha. 2019. "Environmental financialization: What could go wrong?" *Real-World Econ Rev* (87):69-89.

Keswani, A., D. M. Oliver, T. Gutierrez, and R. S. Quilliam. 2016. "Microbial hitchhikers on marine plastic debris: Human exposure risks at bathing waters and beach environments." *Mar Environ Res* 118:10-19. doi: https://doi.org/10.1016/j.marenvres.2016.04.006.

Kiessling, T., L. Gutow, and M. Thiel. 2015. "Marine litter as habitat and dispersal vector." In *Marine Anthropogenic Litter*, edited by M. Bergmann, L. Gutow, and M. Klages, 141-181. Berlin: Springer International.

Kikaki, A., K. Karantzalos, C. A. Power, and D. E. Raitsos. 2020. "Remotely sensing the source and transport of marine plastic debris in Bay Islands of Honduras (Caribbean Sea)." *Remote Sens* 12 (11). doi: 10.3390/rs12111727.

Kilinc, F. S. 2015. "A review of isolation gowns in healthcare: Fabric and gown properties." *J Eng Fibers Fabr* 10 (3):155892501501000313.

Kirstein, I. V., S. Kirmizi, A. Wichels, A. Garin-Fernandez, R. Erler, M. Löder, and G. Gerdts. 2016. "Dangerous hitchhikers? Evidence for potentially pathogenic *Vibrio* spp. on microplastic particles." *Mar Environ Res* 120:1-8. doi: 10.1016/j.marenvres.2016.07.004.

Kleine Jäger, J., and L. Piscicelli. 2021. "Collaborations for circular food packaging: The setup and partner selection process." *Sustain Prod Consum* 26:733-740. doi: https://doi.org/10.1016/j.spc.2020.12.025.

Klosterhaus, S. L., H. M. Stapleton, M. J. La Guardia, and D. J. Greig. 2012. "Brominated and chlorinated flame retardants in San Francisco Bay sediments and wildlife." *Environ Int* 47:56-65. doi: 10.1016/j.envint.2012.06.005.

Knaeps, E., S. Sterckx, G. Strackx, J. Mijnendonckx, M. Moshtaghi, S. P. Garaba, and D. Meire. 2021. "Hyperspectral-reflectance dataset of dry, wet and submerged marine litter." *Earth Sys Sci Data* 13 (2):713-730. doi: 10.5194/essd-13-713-2021.

Knowlton, A. R., and S. D. Kraus. 2001. "Mortality and serious injury of northern right whales (*Eubalaena glacialis*) in the western North Atlantic Ocean." *J. Cetacean Res. Manag*:193-208.

Koelmans, A. A., E. Besseling, and E. M. Foekema. 2014. "Leaching of plastic additives to marine organisms." *Environ Pollut* 187:49-54. doi: 10.1016/j.envpol.2013.12.013.

Koelmans, A. A., E. Besseling, and W. J. Shim. 2015. "Nanoplastics in the aquatic environment. Critical review." In *Marine Anthropogenic Litter*, edited by M. Bergmann, L. Gutow, and M. Klages, 325-340. Berlin: Springer International.

Koelmans, A. A., N. H. Mohamed Nor, E. Hermsen, M. Kooi, S. M. Mintenig, and J. De France. 2019. "Microplastics in freshwaters and drinking water: Critical review and assessment of data quality." *Water Res* 155:410-422. doi: 10.1016/j.watres.2019.02.054.

Kögel, T., Ø. Bjorøy, B. Toto, A. M. Bienfait, and M. Sanden. 2020. "Micro- and nanoplastic toxicity on aquatic life: Determining factors." *Sci Total Environ* 709. doi: 10.1016/j.scitotenv.2019.136050.

Kubota, M. 1994. "A mechanism for the accumulation of floating marine debris north of Hawaii." *J Phys Oceanogr* 24 (5):1059-1064. doi: 10.1175/1520-0485(1994)024<1059:AMFTAO>2.0.CO;2.

Kuczenski, B., C. V. Poulsen, E. L. Gilman, M. Musyl, R. Geyer, and J. Wilson. 2022. "Plastic gear loss estimates from remote observation of industrial fishing activity." *Fish Fish (Oxf)* 23:22–33. doi: 10.1111/faf.12596.

Kühn, S., and J. A. van Franeker. 2020. "Quantitative overview of marine debris ingested by marine megafauna." *Mar Pollut Bull* 151:110858. doi: 10.1016/j.marpolbul.2019.110858.

Kühn, S., E. L. Bravo Rebolledo, and J. A. van Franeker. 2015. "Deleterious effects of litter on marine life." In *Marine Anthropogenic Litter*, edited by M. Bergmann, L. Gutow, and M. Klages, 75-116. Berlin: Springer International.

Kukulka, T., G. Proskurowski, S. Morét-Ferguson, D. W. Meyer, and K. L. Law. 2012. "The effect of wind mixing on the vertical distribution of buoyant plastic debris." *Geophys Res Lett* 39 (7). doi: 10.1029/2012gl051116.

Kumar, R., A. Verma, A. Shome, R. Sinha, S. Sinha, P. K. Jha, R. Kumar, P. Kumar, Shubham, S. Das, P. Sharma, and P. V. V. Prasad. 2021. "Impacts of plastic pollution on ecosystem services, sustainable development goals, and need to focus on circular economy and policy interventions." *Sustainability* 13 (17). doi: 10.3390/su13179963.

Kuroda, M., K. Uchida, T. Tokai, Y. Miyamoto, T. Mukai, K. Imai, K. Shimizu, M. Yagi, Y. Yamanaka, and T. Mituhashi. 2020. "The current state of marine debris on the seafloor in offshore area around Japan." *Mar Pollut Bull* 161 (Pt A):111670. doi: 10.1016/j.marpolbul.2020.111670.

Laist, D. W. 1997. "Impacts of marine debris: Entanglement of marine life in marine debris including a comprehensive list of species with entanglement and ingestion records." In *Marine Debris*, edited by J. M. Coe and D. B. Rogers, 99-140. New York, NY: Springer-Verlag.

Lamb, J. B., B. L. Willis, E. A. Fiorenza, C. S. Couch, R. Howard, D. N. Rader, J. D. True, L. A. Kelly, A. Ahmad, J. Jompa, and C. D. Harvell. 2018. "Plastic waste associated with disease on coral reefs." *Science* 359 (6374):460-462. doi: 10.1126/science.aar3320.

LaMotte, K. R., S. A. Kettenmann, S. N. Munger, and N. J. Waxman. 2021. "States and federal government continue to advance plastics recycling and minimum recycled content mandates." *Nat Law Rev*, June 30.

Lattin, G. L., C. J. Moore, A. F. Zellers, S. L. Moore, and S. B. Weisberg. 2004. "A comparison of neustonic plastic and zooplankton at different depths near the southern California shore." *Mar Pollut Bull* 49 (4):291-294. doi: 10.1016/j.marpolbul.2004.01.020.

Lau, W. W. Y., Y. Shiran, R. M. Bailey, E. Cook, M. R. Stuchtey, J. Koskella, C. A. Velis, L. Godfrey, J. Boucher, M. B. Murphy, R. C. Thompson, E. Jankowska, A. Castillo Castillo, T. D. Pilditch, B. Dixon, L. Koerselman, E. Kosior, E. Favoino, J. Gutberlet, S. Baulch, M. E. Atreya, D. Fischer, K. K. He, M. M. Petit, U. R. Sumaila, E. Neil, M. V. Bernhofen, K. Lawrence, and J. E. Palardy. 2020. "Evaluating scenarios toward zero plastic pollution." *Science* 369 (6510):1455-1461. doi: 10.1126/science.aba9475.

Lavers, J. L., and A. L. Bond. 2017. "Exceptional and rapid accumulation of anthropogenic debris on one of the world's most remote and pristine islands." *Proc Natl Acad Sci USA* 114 (23):6052-6055. doi: 10.1073/pnas.1619818114.

Lavers, J. L., P. B. Sharp, S. Stuckenbrock, and A. L. Bond. 2020. "Entrapment in plastic debris endangers hermit crabs." *J Hazard Mater* 387:121703. doi: 10.1016/j.jhazmat.2019.121703.

Law, K. L. 2017. "Plastics in the marine environment." *Ann Rev Mar Sci* 9:205-229. doi: 10.1146/annurev-marine-010816-060409.

Law, K. L., and R. Narayan. 2022. "Reducing environmental plastic pollution by designing polymer materials for managed end-of-life." *Nat Rev Mater* 7:104–116. doi: 10.1038/s41578-021-00382-0.

Law, K. L., S. Moret-Ferguson, N. A. Maximenko, G. Proskurowski, E. E. Peacock, J. Hafner, and C. M. Reddy. 2010. "Plastic accumulation in the North Atlantic subtropical gyre." *Science* 329 (5996):1185-1188. doi: 10.1126/science.1192321.

Law, K. L., S. E. Moret-Ferguson, D. S. Goodwin, E. R. Zettler, E. Deforce, T. Kukulka, and G. Proskurowski. 2014. "Distribution of surface plastic debris in the eastern Pacific Ocean from an 11-year data set." *Environ Sci Technol* 48 (9):4732-4738. doi: 10.1021/es4053076.

Law, K. L., N. Starr, T. R. Siegel, J. R. Jambeck, N. J. Mallos, and G. H. Leonard. 2020. "The United States' contribution of plastic waste to land and ocean." *Sci Adv* 6. doi: eabd0288.

Le Guen, C., G. Suaria, R. B. Sherley, P. G. Ryan, S. Aliani, L. Boehme, and A. S. Brierley. 2020. "Microplastic study reveals the presence of natural and synthetic fibres in the diet of king penguins (*Aptenodytes patagonicus*) foraging from South Georgia." *Environ Int* 134:105303. doi: 10.1016/j.envint.2019.105303.

Lebreton, L., and A. Andrady. 2019. "Future scenarios of global plastic waste generation and disposal." *Palgrave Commun* 5 (1). doi: 10.1057/s41599-018-0212-7.

Lebreton, L. C. M., J. van der Zwet, J. W. Damsteeg, B. Slat, A. Andrady, and J. Reisser. 2017. "River plastic emissions to the world's oceans." *Nat Commun* 8. doi: 10.1038/ncomms15611.

Lebreton, L., B. Slat, F. Ferrari, B. Sainte-Rose, J. Aitken, R. Marthouse, S. Hajbane, S. Cunsolo, A. Schwarz, A. Levivier, K. Noble, P. Debeljak, H. Maral, R. Schoeneich-Argent, R. Brambini, and J. Reisser. 2018. "Evidence that the Great Pacific Garbage Patch is rapidly accumulating plastic." *Sci Rep* 8 (1). doi: 10.1038/s41598-018-22939-w.

Lee, K.-W., W. J. Shim, O. Y. Kwon, and J.-H. Kang. 2013. "Size-dependent effects of micro polystyrene particles in the marine copepod *Tigriopus japonicus*." *Environ Sci Technol* 47 (19):11278-11283.

Leggett, C., N. Scherer, M. Curry, and R. Bailey. 2014. *Assessing the Economic Benefits of Reductions in Marine Debris: A Pilot Study of Beach Recreation in Orange County, California.* Cambridge, MA: Industrial Economics, Inc.

Leggett, C. G., N. Scherer, T. C. Haab, R. Bailey, J. P. Landrum, and A. Domanski. 2018. "Assessing the economic benefits of reductions in marine debris at Southern California beaches: A random utility travel cost model." *Mar Resour Econ* 33 (2):133-153. doi: 10.1086/697152.

Lenaker, P. L., A. K. Baldwin, S. R. Corsi, S. A. Mason, P. C. Reneau, and J. W. Scott. 2019. "Vertical distribution of microplastics in the water column and surficial sediment from the Milwaukee River Basin to Lake Michigan." *Environ Sci Technol* 53 (21):12227-12237. doi: 10.1021/acs.est.9b03850.

Lenz, R., K. Enders, C. A. Stedmon, D. M. A. Mackenzie, and T. G. Nielsen. 2015. "A critical assessment of visual identification of marine microplastic using Raman spectroscopy for analysis improvement." *Mar Pollut Bull* 100 (1):82-91.

Liboiron, M., F. Liboiron, E. Wells, N. Richard, A. Zahara, C. Mather, H. Bradshaw, and J. Murichi. 2016. "Low plastic ingestion rate in Atlantic cod (*Gadus morhua*) from Newfoundland destined for human consumption collected through citizen science methods." *Mar Pollut Bull* 113 (1-2):428-437. doi: 10.1016/j.marpolbul.2016.10.043.

Lippiatt, S., S. Opfer, and C. Arthur. 2013. "Marine Debris Monitoring and Assessment: Recommendations for Monitoring Debris Trends in the Marine Environment." NOAA Technical Memorandum NOS-OR&R-46.

Liro, M., T. van Emmerik, B. Wyżga, J. Liro, and P. Mikuś. 2020. "Macroplastic storage and remobilization in rivers." *Water* 12 (7). doi: 10.3390/w12072055.

Liu, Y., W. Liu, X. Yang, J. Wang, H. Lin, and Y. Yang. 2021. "Microplastics are a hotspot for antibiotic resistance genes: Progress and perspective." *Sci Total Environ* 773:145643. doi: 10.1016/j.scitotenv.2021.145643.

Lo, H. S., Y. K. Lee, B. H. Po, L. C. Wong, X. Xu, C. F. Wong, C. Y. Wong, N. F. Tam, and S. G. Cheung. 2020. "Impacts of Typhoon Mangkhut in 2018 on the deposition of marine debris and microplastics on beaches in Hong Kong." *Sci Total Environ* 716:137172. doi: 10.1016/j.scitotenv.2020.137172.

Löder, M. G. J., and G. Gerdts. 2015. "Methodology used for the detection and identification of microplastics–a critical appraisal." In *Marine Anthropogenic Litter*, edited by M. Bergmann, L. Gutow, and M. Klages, 201–227. Berlin: Springer International.

Logofet, D. 2021. "Single Use Plastics in China: An Evolving Landscape for Downstream Participants." Herbert Smith Freehills, March 1. https://hsfnotes.com/energy/2021/03/01/single-use-plastics-in-china-an-evolving-landscape-for-downstream-participants.

López-Martínez, S., C. Morales-Caselles, J. Kadar, and M. L. Rivas. 2021. "Overview of global status of plastic presence in marine vertebrates." *Glob Chang Biol* 27 (4):728-737. doi: 10.1111/gcb.15416.

Lorenz, C., L. Roscher, M. S. Meyer, L. Hildebrandt, J. Prume, M. G. J. Löder, S. Primpke, and G. Gerdts. 2019. "Spatial distribution of microplastics in sediments and surface waters of the southern North Sea." *Environ Pollut* 252:1719-1729. doi: 10.1016/j.envpol.2019.06.093.

Lusher, A. L., V. Tirelli, I. O'Connor, and R. Officer. 2015. "Microplastics in Arctic polar waters: The first reported values of particles in surface and sub-surface samples." *Sci Rep* 5. doi: 10.1038/srep14947.

Lusher, A. L., C. O'Donnell, R. Officer, and I. O'Connor. 2016. "Microplastic interactions with North Atlantic mesopelagic fish." *ICES J Mar Sci* 73 (4):1214-1225. doi: 10.1093/icesjms/fsv241.

Lusher, A. L., N. A. Welden, P. Sobral, and M. Cole. 2017. "Sampling, isolating and identifying microplastics ingested by fish and invertebrates." *Anal Meth* 9 (9):1346-1360. doi: 10.1039/c6ay02415g.

Lutz, P. L. "Studies on the ingestion of plastic and latex by sea turtles." 1990. In *Proceedings of the Workshop on the Fate and Impact of Marine Debris, National Marine Fisheries Service, Honolulu*, edited by R. S. Shomura and H. O. Yoshida, pp. 719–735.

Macfayden, G., T. Huntington, and R. Cappell. 2009. *Abandoned, Lost or Otherwise Discarded Fishing Gear*. UNEP Regional Seas Reports and Studies No.185; FAO Fisheries and Aquaculture Technical Paper, No. 523. Rome, UNEP/FAO. Rome: United Nations Environment Programme/Food and Agriculture Organization.

MacLeod, M., H. P. H. Arp, M. B. Tekman, and A. Jahnke. 2021. "The global threat from plastic pollution." *Science* 373 (6550):61-65. doi: 10.1126/science.abg5433.

Mæland, C. E., and R. Staupe-Delgado. 2020. "Can the global problem of marine litter be considered a crisis?" *Risk, Hazards Crisis Pub Policy* 11 (1):87-104.

Mai, L., L.-J. Bao, L. Shi, C. S. Wong, and E. Y. Zeng. 2018. "A review of methods for measuring microplastics in aquatic environments." *Environ Sci Pollut Res* 25 (12):11319-11332.

Mai, L., H. He, L.-J. Bao, L.-Y. Liu, and E. Y. Zeng. 2020. "Plastics are an insignificant carrier of riverine organic pollutants to the coastal oceans." *Environ Sci Technol* 54 (24):15852-15860. doi: 10.1021/acs.est.0c05446.

Manikkam, M., R. Tracey, C. Guerrero-Bosagna, and M. K. Skinner. 2013. "Plastics derived endocrine disruptors (BPA, DEHP and DBP) induce epigenetic transgenerational inheritance of obesity, reproductive disease and sperm epimutations." *PLoS ONE* 8 (1):e55387. doi: 10.1371/journal.pone.0055387.

Marais, M., N. Armitage, and C. Wise. 2004. "The measurement and reduction of urban litter entering stormwater drainage systems: Paper 1-Quantifying the problem using the City of Cape Town as a case study." *Water SA* 30 (4):469-482.

Marine Debris Tracker. 2020. "Marine Debris Tracker Data." https://debristracker.org/data.

Marine Management Organisation. 2016. "Guidance: Marking of Fishing Gear, Retrieval and Notification of Lost Gear." www.gov.uk. https://www.gov.uk/guidance/marking-of-fishing-gear-retrieval-and-notification-of-lost-gear.

Martin, C., S. Parkes, Q. Zhang, X. Zhang, M. F. McCabe, and C. M. Duarte. 2018. "Use of unmanned aerial vehicles for efficient beach litter monitoring." *Mar Pollut Bull* 131 (Pt A):662-673. doi: 10.1016/j.marpolbul.2018.04.045.

Martin, C., E. Corona, G. A. Mahadik, and C. M. Duarte. 2019. "Adhesion to coral surface as a potential sink for marine microplastics." *Environ Pollut* 255. doi: 10.1016/j.envpol.2019.113281.

Martins, K. 2021. "Extended Producer Responsibility Bills Gain Momentum." Retail Industry Leaders Association. https://www.rila.org/blog/2021/04/extended-producer-responsibility-bills.

Martínez-Vicente, V., J. R. Clark, P. Corradi, S. Aliani, M. Arias, M. Bochow, G. Bonnery, M. Cole, A. Cózar, R. Donnelly, F. Echevarría, F. Galgani, S. P. Garaba, L. Goddijn-Murphy, L. Lebreton, H. A. Leslie, P. K. Lindeque, N. Maximenko, F.-R. Martin-Lauzer, D. Moller, P. Murphy, L. Palombi, V. Raimondi, J. Reisser, L. Romero, S. G. H. Simis, S. Sterckx, R. C. Thompson, K. N. Topouzelis, E. van Sebille, J. Mira Veiga, and A. D. Vethaak. 2019. "Measuring marine plastic debris from space: Initial assessment of observation requirements." *Remote Sens* 11 (20). doi: 10.3390/rs11202443.

Masó, M., E. Garcés, F. Pagès, and J. Camp. 2003. "Drifting plastic debris as a potential vector for dispersing Harmful Algal Bloom (HAB) species." *Sci Mar* 67 (1):107-111. doi: 10.3989/scimar.2003.67n1107.

Mato, Y., T. Isobe, H. Takada, H. Kanehiro, C. Ohtake, and T. Kaminuma. 2001. "Plastic resin pellets as a transport medium for toxic chemicals in the marine environment." *Environ Sci Technol* 35 (2):318-324. doi: 10.1021/es0010498.

Matsumura, S., and K. Nasu. 1997. "Distribution of floating debris in the North Pacific Ocean: Sighting surveys 1986–1991." In *Marine Debris*, edited by J. M. Coe and D. B. Rogers, 15-24. New York: Springer.

Matthews, C., F. Moran, and A. K. Jaiswal. 2021. "A review on European Union's strategy for plastics in a circular economy and its impact on food safety." *J Clean Prod* 283. doi: 10.1016/j.jclepro.2020.125263.

Matthews, J. P., L. Ostrovsky, Y. Yoshikawa, S. Komori, and H. Tamura. 2017. "Dynamics and early post-tsunami evolution of floating marine debris near Fukushima Daiichi." *Nat Geosci* 10 (8):598-603. doi: 10.1038/ngeo2975.

Mattsson, K., S. Jocic, I. Doverbratt, and L.-A. Hansson. 2018. "Nanoplastics in the aquatic environment." In *Microplastic Contamination in Aquatic Environments*, edited by E. Y. Zeng, 379-399. Elsevier.

Maximenko, N., P. Corradi, K. L. Law, E. van Sebille, S. P. Garaba, R. S. Lampitt, F. Galgani, V. Martinez-Vicente, L. Goddijn-Murphy, J. Mira Veiga, R. C. Thompson, C. Maes, D. Moller, C. R. Löscher, A. M. Addamo, M. R. Lamson, L. R. Centurioni, N. R. Posth, R. Lumpkin, M. Vinci, A. M. Martins, C. Diogo Pieper, A. Isobe, G. Hanke, M. Edwards, I. P. Chubarenko, E. Rodriguez, S. Aliani, M. Arias, G. P. Asner, A. Brosich, J. T. Carlton, Y. Chao, A.-M. Cook, A. B. Cundy, T. S. Galloway, A. Giorgetti, G. J. Goni, Y. Guichoux, L. E. Haram, B. D. Hardesty, N. Holdsworth, L. Lebreton, H. A. Leslie, I. Macadam-Somer, T. Mace, M. Manuel, R. Marsh, E. Martinez, D. J. Mayor, M. Le Moigne, M. E. M. Jack, M. C. Mowlem, R. W. Obbard, K. Pabortsava, R. Robberson, A.-E. Rotaru, G. M. Ruiz, M. T. Spedicato, M. Thiel, A. Turra, and C. Wilcox. 2019. "Toward the integrated marine debris observing system." *Front Mar Sci* 6. doi: 10.3389/fmars.2019.00447.

McCormick, A. R., T. J. Hoellein, M. G. London, J. Hittie, J. W. Scott, and J. J. Kelly. 2016. "Microplastic in surface waters of urban rivers: Concentration, sources, and associated bacterial assemblages." *Ecosphere* 7 (11). doi: 10.1002/ecs2.1556.

McDermid, K. J., and T. L. McMullen. 2004. "Quantitative analysis of small-plastic debris on beaches in the Hawaiian Archipelago." *Mar Pollut Bull* 48 (7-8):790-794. doi: 10.1016/j.marpolbul.2003.10.017.

McGurty, E. M. 2000. "Warren County, NC, and the emergence of the environmental justice movement: Unlikely coalitions and shared meanings in local collective action." *Soc Nat Resour* 13 (4):373-387. doi: 10.1080/089419200279027.

McWilliams, M., M. Liboiron, and Y. Wiersma. 2018. "Rocky shoreline protocols miss microplastics in marine debris surveys (Fogo Island, Newfoundland and Labrador)." *Mar Pollut Bull* 129 (2):480-486. doi: 10.1016/j.marpolbul.2017.10.018.

Meijer, L. J. J., T. van Emmerik, R. van der Ent, C. Schmidt, and L. Lebreton. 2021. "More than 1000 rivers account for 80% of global riverine plastic emissions into the ocean." *Sci Adv* 7 (18). doi: 10.1126/sciadv.aaz5803.

Merrell, T. R. 1980. "Accumulation of plastic litter on beaches of Amchitka Island, Alaska." *Mar Environ Res* 3 (3):171-184. doi: 10.1016/0141-1136(80)90025-2.

Mezzacapo, M., M. J. Donohue, C. Smith, A. El-Kadi, K. Falinski, and D. T. Lerner. 2021. "Review article: Hawai'i's cesspool problem: Review and recommendations for water resources and human health." *J Contemp Water Res Educ* 170 (1):35-75. doi: 10.1111/j.1936-704X.2020.03339.x.

Millet, H., P. Vangheluwe, C. Block, A. Sevenster, L. Garcia, and R. Antonopoulos. 2018. "The nature of plastics and their societal usage." In *Plastics and the Environment*, edited by R. M. Harrison and R. E. Hester, 1-20. Royal Society of Chemistry.

Ministry of Environment, Japan. 2021. *G20 Report on Actions Against Marine Plastic Litter: Third Information Sharing Based on the G20 Implementation Framework*. Tokyo, Ministry of the Environment, Japan. https://g20mpl.org/wp-content/uploads/2021/07/G20MPL_report_2021.pdf.

Miralles, L., M. Gomez-Agenjo, F. Rayon-Viña, G. Gyraite·, and E. Garcia-Vazquez. 2018. "Alert calling in port areas: Marine litter as possible secondary dispersal vector for hitchhiking invasive species." *J Nat Conserv* 42:12-18. doi: 10.1016/j.jnc.2018.01.005.

Mizutani, J. 2018. "In the backyard of segregated neighborhoods: An environmental justice case study of Louisiana." *Georgetown Environ Law Rev* 31:363.

Mongelluzzo, B. 2018. US export recyclables — China shifts, market shudders. J Commerce Online

Moore, C. J. 2008. "Synthetic polymers in the marine environment: A rapidly increasing, long-term threat." *Environ Res* 108 (2):131-139. doi: 10.1016/j.envres.2008.07.025.

Moore, C. J., G. L. Lattin, and A. F. Zellers. 2011. "Quantity and type of plastic debris flowing from two urban rivers to coastal waters and beaches of Southern California." *J Integr Coastal Zone Manag* 11 (1):65-73.

Moore, M. J., A. Bogomolni, R. Bowman, P. K. Hamilton, C. T. Harry, A. R. Knowlton, S. Landry, D. S. Rotstein, and K. Touhey. 2006. "Fatally entangled right whales can die extremely slowly." In *OCEANS 2006 Conference Proceedings*. IEEE. doi: 10.1109/OCEANS.2006.306792.

Moore, M. J., T. K. Rowles, D. A. Fauquier, J. D. Baker, I. Biedron, J. W. Durban, P. K. Hamilton, A. G. Henry, A. R. Knowlton, W. A. McLellan, C. A. Miller, R. M. Pace, 3rd, H. M. Pettis, S. Raverty, R. M. Rolland, R. S. Schick, S. M. Sharp, C. R. Smith, L. Thomas, J. M. V. der Hoop, and M. H. Ziccardi. 2021. "REVIEW: Assessing North Atlantic right whale health: Threats, and development of tools critical for conservation of the species." *Dis Aquat Organ* 143:205-226. doi: 10.3354/dao03578.

Moore, S., T. Hale, S. B. Weisberg, L. Flores, and P. Kauhanen. 2021. *California Trash Monitoring Methods and Assessments Playbook*. Richmond, California: San Francisco Estuary Institute.

Morishige, C., M. J. Donohue, E. Flint, C. Swenson, and C. Woolaway. 2007. "Factors affecting marine debris deposition at French Frigate Shoals, northwestern Hawaiian Islands Marine National Monument, 1990-2006." *Mar Pollut Bull* 54 (8):1162-1169. doi: 10.1016/j.marpolbul.2007.04.014.

Moshtaghi, M., E. Knaeps, S. Sterckx, S. Garaba, and D. Meire. 2021. "Spectral reflectance of marine macroplastics in the VNIR and SWIR measured in a controlled environment." *Sci Rep* 11 (1):5436. doi: 10.1038/s41598-021-84867-6.

Moy, K., B. Neilson, A. Chung, A. Meadows, M. Castrence, S. Ambagis, and K. Davidson. 2018. "Mapping coastal marine debris using aerial imagery and spatial analysis." *Mar Pollut Bull* 132:52-59. doi: 10.1016/j.marpolbul.2017.11.045.

Mrosovsky, N., G. D. Ryan, and M. C. James. 2009. "Leatherback turtles: The menace of plastic." *Mar Pollut Bull* 58 (2):287-289. doi: 10.1016/j.marpolbul.2008.10.018.

MSFD Technical Subgroup on Marine Litter. 2013. Guidance on Monitoring of Marine Litter in European Seas: A guidance document within the Common Implementation Strategy for the Marine Strategy Framework Directive. Luxembourg: Joint Research Centre.

Munno, K., P. A. Helm, C. Rochman, T. George, and D. A. Jackson. 2021. "Microplastic contamination in Great Lakes fish." *Conserv Biol*. doi: 10.1111/cobi.13794.

Murray, C. C., N. Maximenko, and S. Lippiatt. 2018. "The influx of marine debris from the Great Japan Tsunami of 2011 to North American shorelines." *Mar Pollut Bull* 132:26-32. doi: 10.1016/j.marpolbul.2018.01.004.

Murray, F., and P. R. Cowie. 2011. "Plastic contamination in the decapod crustacean *Nephrops norvegicus* (Linnaeus, 1758)." *Mar Pollut Bull* 62 (6):1207-1217. doi: 10.1016/j.marpolbul.2011.03.032.

Myers, H. J., and M. J. Moore. 2020. "Reducing effort in the U.S. American lobster (*Homarus americanus*) fishery to prevent North Atlantic right whale (*Eubalaena glacialis*) entanglements may support higher profits and long-term sustainability." *Mar Policy* 118. doi: 10.1016/j.marpol.2020.104017.

Myers, H. J., M. J. Moore, M. F. Baumgartner, S. W. Brillant, S. K. Katona, A. R. Knowlton, L. Morissette, H. M. Pettis, G. Shester, and T. B. Werner. 2019. "Ropeless fishing to prevent large whale entanglements: Ropeless Consortium report." *Mar Policy* 107. doi: 10.1016/j.marpol.2019.103587.

Naji, A., M. Nuri, and A. D. Vethaak. 2018. "Microplastics contamination in molluscs from the northern part of the Persian Gulf." *Environl Pollut* 235:113-120. doi: 10.1016/j.envpol.2017.12.046.

NASEM (National Academies of Sciences, Engineering, and Medicine). 2020. *Closing the Loop on the Plastics Dilemma: Proceedings of a Workshop–in Brief*, edited by A. F. Johnson and E. Mantus. Washington, DC: The National Academies Press.

National Geographic. 2021. *Sea to Source Summary Report*. Washington, D.C.

NOAA Marine Debris Program. 2020a. *Marine Debris Program Accomplishments Report*. Silver Spring, MD: National Oceanic and Atmospheric Administration.

NOAA Marine Debris Program. 2020b. *NOAA Marine Debris Program FY 2021-2025 Strategic Plan*. Silver Spring, MD: National Oceanic and Atmospheric Administration.

NOAA Marine Debris Program. 2021a. "Funding Opportunities." https://marinedebris.noaa.gov/resources/funding-opportunities.

NOAA Marine Debris Program. 2021b. "Reports and Technical Memos." https://marine debris.noaa.gov/reports-and-technical-memos.

National Research Council. 1995. *Clean Ships, Clean Ports, Clean Oceans: Controlling Garbage and Plastic Wastes at Sea*. Washington, DC: The National Academies Press.

National Research Council. 2009. *Tackling Marine Debris in the 21st Century*. Washington, D.C.: The National Academies Press.

NDRC and MEE (National Development and Reform Commission and the Ministry of Ecology and Environment). 2020. "Opinions on Further Strengthening the Treatment of Plastic Pollution."

Nelms, S. E., T. S. Galloway, B. J. Godley, D. S. Jarvis, and P. K. Lindeque. 2018. "Investigating microplastic trophic transfer in marine top predators." *Environ Pollut* 238:999-1007. doi: 10.1016/j.envpol.2018.02.016.

Nihei, Y., T. Yoshida, T. Kataoka, and R. Ogata. 2020. "High-resolution mapping of Japanese microplastic and macroplastic emissions from the land into the sea." *Water* 12 (4). doi: 10.3390/w12040951.

Nizzetto, L., G. Bussi, M. N. Futter, D. Butterfield, and P. G. Whitehead. 2016. "A theoretical assessment of microplastic transport in river catchments and their retention by soils and river sediments." *Environ Sci Process Impacts* 18 (8):1050-1059. doi: 10.1039/c6em00206d.

North, E. J., and R. U. Halden. 2013. "Plastics and environmental health: The road ahead." *Rev Environ Health* 28 (1):1-8. doi: 10.1515/reveh-2012-0030.

Oberbeckmann, S., and M. Labrenz. 2020. "Marine microbial assemblages on microplastics: Diversity, adaptation, and role in degradation." *Ann Rev Mar Sci* 12:209-232. doi: 10.1146/annurev-marine-010419-010633.

Ocean Conservancy. 2021. *Pandemic Pollution: The Rising Tide of Plastic.* https://ocean conservancy.org/trash-free-seas/take-deep-dive/pandemic-pollution-rising-tide-plastic-ppe.

Ocean Outcomes. 2020. *Ghost Gear Legislation Analysis.* Gland, Switzerland: World Wide Fund For Nature (Formerly World Wildlife Fund).

OECD (Organisation for Economic Co-operation and Development). 2018. *Improving Markets for Recycled Plastics.* Paris: OECD Publishing. http://dx.doi.org/10.1787/9789264301016-en%0D.

Opfer, S., C. Arthur, and S. Lippiatt. 2012. "NOAA Marine Debris Shoreline Survey Field Guide." Washington, D.C.: NOAA.

O'Rourke, D., and S. Connolly. 2003. "Justoil? The distribution of environmental and social impacts of oil production and consumption." *Annu Rev Environ Resourc* 28 (1):587-617. doi: 10.1146/annurev.energy.28.050302.105617.

Ostle, C., R. C. Thompson, D. Broughton, L. Gregory, M. Wootton, and D. G. Johns. 2019. "The rise in ocean plastics evidenced from a 60-year time series." *Nat Commun* 10 (1):1622. doi: 10.1038/s41467-019-09506-1.

Paben, J. 2021. "British Columbia EPR program hits 85% recovery rate." Resource Recycling. https://resource-recycling.com/recycling/2021/07/20/british-columbia-epr-program-hits-85-recovery-rate.

Pallister, C. 2012. *Southern Montague Island 2012 Marine Debris Cleanup Final Report.* Juneau, AK: Marine Conservation Alliance Foundation.

Panno, S. V., W. R. Kelly, J. Scott, W. Zheng, R. E. McNeish, N. Holm, T. J. Hoellein, and E. L. Baranski. 2019. "Microplastic contamination in karst groundwater systems." *Ground Water* 57 (2):189-196. doi: 10.1111/gwat.12862.

Pasternak, G., C. A. Ribic, E. Spanier, and D. Zviely. 2021. "Stormwater systems as a source of marine debris: A case study from the Mediterranean coast of Israel." *J Coast Conserv* 25 (1). doi: 10.1007/s11852-021-00818-3.

Patrício Silva, A. L., J. C. Prata, T. R. Walker, A. C. Duarte, W. Ouyang, D. Barcelò, and T. Rocha-Santos. 2021. "Increased plastic pollution due to COVID-19 pandemic: Challenges and recommendations." *Chem Eng J* 405:126683-126683. doi: 10.1016/j.cej.2020.126683.

Pauly, D., and D. Zeller. 2016. "Catch reconstructions reveal that global marine fisheries catches are higher than reported and declining." *Nat Commun* 7. doi: 10.1038/ncomms10244.

Peng, L., D. Fu, H. Qi, C. Q. Lan, H. Yu, and C. Ge. 2020. "Micro- and nano-plastics in marine environment: Source, distribution and threats — A review." *Sci Total Environ* 698:134254. doi: https://doi.org/10.1016/j.scitotenv.2019.134254.

Pérez-Guevara, F., G. Kutralam-Muniasamy, and V. C. Shruti. 2021. "Critical review on microplastics in fecal matter: Research progress, analytical methods and future outlook." *Sci Total Environ* 778:146395. doi: 10.1016/j.scitotenv.2021.146395.

Perez-Venegas, D. J., M. Seguel, H. Pavés, J. Pulgar, M. Urbina, C. Ahrendt, and C. Galbán-Malagón. 2018. "First detection of plastic microfibers in a wild population of South American fur seals (*Arctocephalus australis*) in the Chilean Northern Patagonia." *Mar Pollut Bull* 136:50-54. doi: 10.1016/j.marpolbul.2018.08.065.

PET Resin Association. 2015. "PET by the Numbers." http://www.petresin.org/news_PET bythenumbers.asp.

Pettit, T. N., G. S. Grant, and G. C. Whittow. 1981. "Ingestion of plastics by Laysan albatross." *The Auk* 98 (4):839-841.

Pham, D. N., L. Clark, and M. Li. 2021. "Microplastics as hubs enriching antibiotic-resistant bacteria and pathogens in municipal activated sludge." *J Hazard Mater Lett* 2. doi: 10.1016/j.hazl.2021.100014.

Piccardo, M., M. Renzi, and A. Terlizzi. 2020. "Nanoplastics in the oceans: Theory, experimental evidence and real world." *Mar Pollut Bull* 157:111317. doi: 10.1016/j.marpolbul.2020.111317.

Pichel, W. G., T. S. Veenstra, J. H. Churnside, E. Arabini, K. S. Friedman, D. G. Foley, R. E. Brainard, D. Kiefer, S. Ogle, P. Clemente-Colon, and X. Li. 2012. "GhostNet marine debris survey in the Gulf of Alaska—satellite guidance and aircraft observations." *Mar Pollut Bull* 65 (1-3):28-41. doi: 10.1016/j.marpolbul.2011.10.009.

Pilz, H., B. Brandt, and R. Fehringer. 2010. *The Impact of Plastics on Life Cycle Energy Consumption and Greenhouse Gas Emissions in Europe.* Denkstatt.

Plastics Europe. 2020. Plastics – the Facts 2020: An Analysis of European Plastics Production, Demand and Waste Data.

Polasek, L., J. Bering, H. Kim, P. Neitlich, B. Pister, M. Terwilliger, K. Nicolato, C. Turner, and T. Jones. 2017. "Marine debris in five national parks in Alaska." *Mar Pollut Bull* 117 (1-2):371-379. doi: 10.1016/j.marpolbul.2017.01.085.

Powell, J. T., and M. R. Chertow. 2019. "Quantity, components, and value of waste materials landfilled in the United States." *Journal of Industrial Ecology* 23 (2):466-479. doi: 10.1111/jiec.12752.

Primpke, S., S. H. Christiansen, W. Cowger, H. De Frond, A. Deshpande, M. Fischer, E. B. Holland, M. Meyns, B. A. O'Donnell, and B. E. Ossmann. 2020. "Critical assessment of analytical methods for the harmonized and cost-efficient analysis of microplastics." *Appl Spectrosc* 74 (9):1012-1047.

Provencher, J. F., A. J. Gaston, M. L. Mallory, P. D. O'Hara, and H. G. Gilchrist. 2010. "Ingested plastic in a diving seabird, the thick-billed murre (*Uria lomvia*), in the eastern Canadian Arctic." *Mar Pollut Bull* 60 (9):1406-11. doi: 10.1016/j.marpolbul.2010.05.017.

Public Health Law Center. 2020. "Tobacco Product Waste: Frequently Asked Questions." https://www.publichealthlawcenter.org/sites/default/files/resources/Tobacco-Product-Waste-CA-FAQ.pdf.

Quinn, M. 2021. "Washington Governor Signs Minimum Recycled Content Bill." *Waste Dive.* https://www.wastedive.com/news/minimum-recycled-content-bill-washington-inslee/598614/.

Ragaert, K., L. Delva, and K. Van Geem. 2017. "Mechanical and chemical recycling of solid plastic waste." *Waste Manag* 69:24-58. doi: 10.1016/j.wasman.2017.07.044.

Ragusa, A., A. Svelato, C. Santacroce, P. Catalano, V. Notarstefano, O. Carnevali, F. Papa, M. C. A. Rongioletti, F. Baiocco, S. Draghi, E. D'Amore, D. Rinaldo, M. Matta, and E. Giorgini. 2021. "Plasticenta: First evidence of microplastics in human placenta." *Environ Int* 146:106274. doi: 10.1016/j.envint.2020.106274.

Rech, S., S. Salmina, Y. J. Borrell Pichs, and E. García-Vazquez. 2018. "Dispersal of alien invasive species on anthropogenic litter from European mariculture areas." *Mar Pollut Bull* 131 (Pt A):10-16. doi: 10.1016/j.marpolbul.2018.03.038.

Redling, A. 2021. "Waste and Recycling Industry Projections for 2021." *Waste Today*. January 27.

Reed, S., M. Clark, R. Thompson, and K. A. Hughes. 2018. "Microplastics in marine sediments near Rothera Research Station, Antarctica." *Mar Pollut Bull* 133:460-463. doi: 10.1016/j.marpolbul.2018.05.068.

Reichert, J., J. Schellenberg, P. Schubert, and T. Wilke. 2018. "Responses of reef building corals to microplastic exposure." *Environ Pollut* 237:955-960. doi: 10.1016/j.envpol.2017.11.006.

Reisser, J., B. Slat, K. Noble, K. du Plessis, M. Epp, M. Proietti, J. de Sonneville, T. Becker, and C. Pattiaratchi. 2015. "The vertical distribution of buoyant plastics at sea: An observational study in the North Atlantic Gyre." *Biogeosciences* 12 (4):1249-1256. doi: 10.5194/bg-12-1249-2015.

Rey, S. F., J. Franklin, and S. J. Rey. 2021. "Microplastic pollution on island beaches, Oahu, Hawai`i." *PLoS ONE* 16 (2):e0247224. doi: 10.1371/journal.pone.0247224.

Ribic, C. A., S. W. Johnson, and C. A. Cole. 1997. "Distribution, type, accumulation, and source of marine debris in the US 1989–1993." In *Marine Debris*, edited by J. M. Coe and D. B. Rogers, 35–47. New York, NY: Springer-Verlag.

Ribic, C. A., S. B. Sheavly, and J. Klavitter. 2012. "Baseline for beached marine debris on Sand Island, Midway Atoll." *Mar Pollut Bull* 64 (8):1726-1729. doi: 10.1016/j.marpolbul.2012.04.001.

Ribic, C. A., S. B. Sheavly, D. J. Rugg, and E. S. Erdmann. 2010. "Trends and drivers of marine debris on the Atlantic coast of the United States 1997-2007." *Mar Pollut Bull* 60 (8):1231-1242. doi: 10.1016/j.marpolbul.2010.03.021.

Richardson, K., B. D. Hardesty, and C. Wilcox. 2019. "Estimates of fishing gear loss rates at a global scale: A literature review and meta-analysis." *Fish Fish (Oxf)* 20 (6):1218-1231. doi: 10.1111/faf.12407.

Richardson, K., C. Wilcox, J. Vince, and B. D. Hardesty. 2021. "Challenges and misperceptions around global fishing gear loss estimates." *Mar Policy* 129. doi: 10.1016/j.marpol.2021.104522.

Rios, L. M., C. Moore, and P. R. Jones. 2007. "Persistent organic pollutants carried by synthetic polymers in the ocean environment." *Mar Pollut Bull* 54 (8):1230-1237. doi: 10.1016/j.marpolbul.2007.03.022.

Roberge, R. J. 2016. "Face shields for infection control: A review." *J Occup Environ Hyg* 13 (4):235-242. doi: 10.1080/15459624.2015.1095302.

Rocha, R. J. M., A. C. M. Rodrigues, D. Campos, L. H. Cícero, A. P. L. Costa, D. A. M. Silva, M. Oliveira, A. M. V. M. Soares, and A. L. Patrício Silva. 2020. "Do microplastics affect the zoanthid *Zoanthus sociatus*?" *Sci Total Environ* 713:136659. doi: 10.1016/j.scitotenv.2020.136659.

Rochman, C. M., E. Hoh, B. T. Hentschel, and S. Kaye. 2013. "Long-term field measurement of sorption of organic contaminants to five types of plastic pellets: Implications for plastic marine debris." *Environ Sci Technol* 47 (3):1646-54. doi: 10.1021/es303700s.

Rochman, C. M., B. T. Hentschel, and S. J. Teh. 2014. "Long-term sorption of metals is similar among plastic types: Implications for plastic debris in aquatic environments." *PLoS ONE* 9 (1):e85433. doi: 10.1371/journal.pone.0085433.

Rochman, C. M., A. Tahir, S. L. Williams, D. V. Baxa, R. Lam, J. T. Miller, F. C. Teh, S. Werorilangi, and S. J. Teh. 2015. "Anthropogenic debris in seafood: Plastic debris and fibers from textiles in fish and bivalves sold for human consumption." *Sci Rep* 5:14340. doi: 10.1038/srep14340.

Rochman, C. M., F. Regan, and R. C. Thompson. 2017. "On the harmonization of methods for measuring the occurrence, fate and effects of microplastics." *Anal Meth* 9 (9):1324-1325.

Rogoff, M. J., and D. E. Ross. 2016. "The future of recycling in the United States." *Waste Manag Res* 34 (3):181-183. doi: 10.1177/0734242X16629599.

Rorrer, J. E., G. T. Beckham, and Y. Roman-Leshkov. 2021. "Conversion of polyolefin waste to liquid alkanes with Ru-based catalysts under mild conditions." *JACS Au* 1 (1):8-12. doi: 10.1021/jacsau.0c00041.

Ryan, P. G. 1987. "The incidence and characteristics of plastic particles ingested by seabirds." *Mar Environ Res* 23 (3):175-206. doi: 10.1016/0141-1136(87)90028-6.

Ryan, P. G. 1988. "Intraspecific variation in plastic ingestion by seabirds and the flux of plastic through seabird populations." *The Condor* 90 (2):446-452. doi: 10.2307/1368572.

Ryan, P. G., and M. W. Fraser. 1988. "The use of great skua pellets as indicators of plastic pollution in seabirds." *Emu* 88 (1):16-19.

Ryan, P. G., and C. L. Moloney. 1993. "Marine litter keeps increasing." *Nature* 361 (6407):23. doi: 10.1038/361023a0.

Ryan, P. G., C. J. Moore, J. A. van Franeker, and C. L. Moloney. 2009. "Monitoring the abundance of plastic debris in the marine environment." *Philos Trans R Soc Lond B Biol Sci* 364 (1526):1999-2012. doi: 10.1098/rstb.2008.0207.

Ryan, P. G., A. Lamprecht, D. Swanepoel, and C. L. Moloney. 2014. "The effect of fine-scale sampling frequency on estimates of beach litter accumulation." *Mar Pollut Bull* 88 (1-2):249-254. doi: 10.1016/j.marpolbul.2014.08.036.

Saha, R., and P. Mohai. 2005. "Historical context and hazardous waste facility siting: Understanding temporal patterns in Michigan." *Soc Probl* 52 (4):618-648. doi: 10.1525/sp.2005.52.4.618.

Saliu, F., S. Montano, B. Leoni, M. Lasagni, and P. Galli. 2019. "Microplastics as a threat to coral reef environments: Detection of phthalate esters in neuston and scleractinian corals from the Faafu Atoll, Maldives." *Mar Pollut Bull* 142:234-241. doi: 10.1016/j.marpolbul.2019.03.043.

Saliu, F., M. Veronelli, C. Raguso, D. Barana, P. Galli, and M. Lasagni. 2021. "The release process of microfibers: From surgical face masks into the marine environment." *Environ Adv* 4. doi: 10.1016/j.envadv.2021.100042.

Sandra, M., L. Devriese, F. De Raedemaecker, B. Lonneville, I. Lukic, S. Altvater, M. Compa Ferrer, S. Deudero, B. Torres Hansjosten, C. Alomar Mascaró, I. Gin, M. Vale, M. Zorgno, and M. Mata Lara. 2020. *Knowledge Wave on Marine Litter from Aquaculture Sources.* D2.2 Aqua-Lit project. Oostende, Belgium.

Santana-Viera, S., S. Montesdeoca-Esponda, R. Guedes-Alonso, Z. Sosa-Ferrera, and J. J. Santana-Rodríguez. 2021. "Organic pollutants adsorbed on microplastics: Analytical methodologies and occurrence in oceans." *Trends Environ Anal Chem* 29. doi: 10.1016/j.teac.2021.e00114.

Santos, R. G., G. E. Machovsky-Capuska, and R. Andrades. 2021. "Plastic ingestion as an evolutionary trap: Toward a holistic understanding." *Science* 373 (6550):56-60. doi: 10.1126/science.abh0945.

Savinelli, B., T. Vega Fernández, N. M. Galasso, G. D'Anna, C. Pipitone, F. Prada, A. Zenone, F. Badalamenti, and L. Musco. 2020. "Microplastics impair the feeding performance of a Mediterranean habitat-forming coral." *Mar Environ Res* 155. doi: 10.1016/j.marenvres.2020.104887.

Savoca, M. S., M. E. Wohlfeil, S. E. Ebeler, and G. A. Nevitt. 2016. "Marine plastic debris emits a keystone infochemical for olfactory foraging seabirds." *Sci Adv* 2 (11):e1600395. doi: 10.1126/sciadv.1600395.

Schmidt, C., T. Krauth, and S. Wagner. 2017. "Export of plastic debris by rivers into the sea." *Environ Sci Technol* 51 (21):12246-12253. doi: 10.1021/acs.est.7b02368.

Schuyler, Q., B. D. Hardesty, C. Wilcox, and K. Townsend. 2012. "To eat or not to eat? Debris selectivity by marine turtles." *PLoS ONE* 7 (7):e40884. doi: 10.1371/journal.pone.0040884.

Schuyler, Q., B. D. Hardesty, C. Wilcox, and K. Townsend. 2014. "Global analysis of anthropogenic debris ingestion by sea turtles." *Conserv Biol* 28 (1):129-139. doi: 10.1111/cobi.12126.

Schuyler, Q., B. D. Hardesty, T. J. Lawson, K. Opie, and C. Wilcox. 2018. "Economic incentives reduce plastic inputs to the ocean." *Mar Policy* 96:250-255. doi: 10.1016/j.marpol.2018.02.009.

Schuyler, Q., C. Wilcox, T. J. Lawson, R. R. M. K. P. Ranatunga, C. S. Hu, and B. D. Hardesty. 2021. "Human population density is a poor predictor of debris in the environment." *Front Environ Sci* 9. doi: 10.3389/fenvs.2021.583454.

Schwarz, A. E., T. N. Ligthart, E. Boukris, and T. van Harmelen. 2019. "Sources, transport, and accumulation of different types of plastic litter in aquatic environments: A review study." *Mar Pollut Bull* 143:92-100. doi: 10.1016/j.marpolbul.2019.04.029.

SCVURPPP (Santa Clara Valley Urban Runoff Pollution Prevention Program). 2021. "Trash Sources in the Santa Clara Valley." https://scvurppp.org/trash/sources/.

Sharma, R., and G. Ghoshal. 2018. "Emerging trends in food packaging." *Nutr Food Sci* 48 (5):764-779. doi: 10.1108/nfs-02-2018-0051.

Sheavly, S. B. 2007. *National Marine Debris Monitoring Program: Final Program Report, Data Analysis Summary.* Prepared for U.S. Environmental Protection Agency by Ocean Conservancy.

Shen, M., S. Ye, G. Zeng, Y. Zhang, L. Xing, W. Tang, X. Wen, and S. Liu. 2020. "Can microplastics pose a threat to ocean carbon sequestration?" *Mar Pollut Bull* 150. doi: 10.1016/j.marpolbul.2019.110712.

Shi, C., Z.-C. Li, L. Caporaso, L. Cavallo, L. Falivene, and E. Y. X. Chen. 2021. "Hybrid monomer design for unifying conflicting polymerizability, recyclability, and performance properties." *Chem* 7 (3):670-685. doi: 10.1016/j.chempr.2021.02.003.

Shomura, R. S., and H. O. Yoshida. 1985. *Proceedings of the Workshop on the Fate and Impact of Marine Debris, 27-29 November 1984, Honolulu, Hawaii.* Sea grant cooperative report 85-04, NOAA-TM-NMFS-SWFC. National Oceanic and Atmospheric Administration.

Silva, A. B., A. S. Bastos, C. I. L. Justino, J. P. da Costa, A. C. Duarte, and T. A. P. Rocha-Santos. 2018. "Microplastics in the environment: Challenges in analytical chemistry - A review." *Anal Chim Acta* 1017:1-19. doi: 10.1016/j.aca.2018.02.043.

Singh, S., and S. S.-L. Li. 2012. "Epigenetic effects of environmental chemicals bisphenol A and phthalates." *Int J Mol Sci* 13 (8):10143-10153.

Skoczinski, P., M. Carus, D. de Guzman, H. Käb, R. Chinthapalli, J. Ravenstijn, W. Baltus, and A. Raschka. 2021. *Bio-based Building Blocks and Polymers – Global Capacities, Production and Trends 2020 – 2025.* Hürth, Germany: nova-Institute.

Smith, M., D. C. Love, C. M. Rochman, and R. A. Neff. 2018. "Microplastics in seafood and the implications for human health." *Curr Environ Health Rep* 5 (3):375-386. doi: 10.1007/s40572-018-0206-z.

Smith, S. D. A., and A. Markic. 2013. "Estimates of marine debris accumulation on beaches are strongly affected by the temporal scale of sampling." *PLoS ONE* 8 (12). doi: 10.1371/journal.pone.0083694.

Song, Z., X. Yang, F. Chen, F. Zhao, Y. Zhao, L. Ruan, Y. Wang, and Y. Yang. 2019. "Fate and transport of nanoplastics in complex natural aquifer media: Effect of particle size and surface functionalization." *Sci Total Environ* 669:120-128. doi: 10.1016/j.scitotenv.2019.03.102.

Staub, C. 2021. "Paper and plastic exports drop again in 2020." Resource Recycling. https://resource-recycling.com/recycling/2021/02/05/paper-and-plastic-exports-drop-again-in-2020.

Steffen, L. 2020. "Norway Leads The World With Its Incredible Recycling Scheme." Intelligent Living, May 24.

Stickel, B. H., A. Jahn, and W. Kier. 2012. *The Cost to West Coast Communities of Dealing with Trash, Reducing Marine Debris.* Kier Associates.

Sun, X., J. Liang, M. Zhu, Y. Zhao, and B. Zhang. 2018. "Microplastics in seawater and zooplankton from the Yellow Sea." *Environ Pollut* 242:585-595. doi: https://doi.org/10.1016/j.envpol.2018.07.014.

Sutton, R., D. Lin, M. Sedlak, C. Box, A. Gilbreath, R. Holleman, L. Miller, A. Wong, K. Munno, X. Zhu, and C. Rochman. 2019. *Understanding Microplastic Levels, Pathways, and Transport in the San Francisco Bay Region*. SFEI Contribution No. 950. Richmond, CA: San Francisco Estuary Institute.

Tagg, A. S., M. Sapp, J. P. Harrison, and J. J. Ojeda. 2015. "Identification and quantification of microplastics in wastewater using focal plane array-based reflectance micro-FT-IR imaging." *Anal Chem* 87 (12):6032-6040.

Tasseron, P., T. van Emmerik, J. Peller, L. Schreyers, and L. Biermann. 2021. "Advancing floating macroplastic detection from space using experimental hyperspectral imagery." *Remote Sens* 13 (12). doi: 10.3390/rs13122335.

ter Halle, A., L. Ladirat, X. Gendre, D. Goudouneche, C. Pusineri, C. Routaboul, C. Tenailleau, B. Duployer, and E. Perez. 2016. "Understanding the fragmentation pattern of marine plastic debris." *Environ Sci Technol* 50 (11):5668-5675. doi: 10.1021/acs.est.6b00594.

Ternes, M. E. and S. Fulton, S. 2020. Overview of United States Law Governing Solid and Water Waste Management. Presentation at an online meeting of the Committee on the U.S. Contributions to Global Ocean Plastic Waste, December 7.

Teuten, E. L., S. J. Rowland, T. S. Galloway, and R. C. Thompson. 2007. "Potential for plastics to transport hydrophobic contaminants." *Environ Sci Technol* 41 (22):7759-7764. doi: 10.1021/es071737s.

Themistocleous, K., C. Papoutsa, S. Michaelides, and D. Hadjimitsis. 2020. "Investigating detection of floating plastic litter from space using Sentinel-2 imagery." *Remote Sens* 12 (16). doi: 10.3390/rs12162648.

Thiel, M., I. A. Hinojosa, L. Miranda, J. F. Pantoja, M. M. Rivadeneira, and N. Vasquez. 2013. "Anthropogenic marine debris in the coastal environment: A multi-year comparison between coastal waters and local shores." *Mar Pollut Bull* 71 (1-2):307-316. doi: 10.1016/j.marpolbul.2013.01.005.

Thiel, M., S. Hong, J. R. Jambeck, M. Gatta-Rosemary, D. Honorato-Zimmer, T. Kiessling, K. Knickmeier, and K. Kruse. 2017. "Marine litter – bringing together citizen scientists from around the world." In *Citizen Science for Coastal and Marine Conservation*, edited by J. A. Cigliano and H. L Ballard, 104-131. Routledge.

Thompson, R. C., Y. Olsen, R. P. Mitchell, A. Davis, S. J. Rowland, A. W. John, D. McGonigle, and A. E. Russell. 2004. "Lost at sea: Where is all the plastic?" *Science* 304 (5672):838. doi: 10.1126/science.1094559.

Thompson, R. C., S. H. Swan, C. J. Moore, and F. S. vom Saal. 2009. "Our plastic age." *Philos Trans R Soc Lond B Biol Sci* 364 (1526):1973-1976. doi: 10.1098/rstb.2009.0054.

Tian, Z., H. Zhao, K. T. Peter, M. Gonzalez, J. Wetzel, C. Wu, X. Hu, J. Prat, E. Mudrock, R. Hettinger, A. E. Cortina, R. G. Biswas, F. V. C. Kock, R. Soong, A. Jenne, B. Du, F. Hou, H. He, R. Lundeen, A. Gilbreath, R. Sutton, N. L. Scholz, J. W. Davis, M. C. Dodd, A. Simpson, J. K. McIntyre, and E. P. Kolodziej. 2021. "A ubiquitous tire rubber-derived chemical induces acute mortality in coho salmon." *Science* 371 (6525):185-189. doi: 10.1126/science.abd6951.

Tibbetts, J., S. Krause, I. Lynch, and G. Sambrook Smith. 2018. "Abundance, distribution, and drivers of microplastic contamination in urban river environments." *Water* 10 (11). doi: 10.3390/w10111597.

Tishman Environment and Design Center. 2019. *U.S. Municipal Solid Waste Incinerators: An Industry in Decline*.

Topouzelis, K., A. Papakonstantinou, and S. P. Garaba. 2019. "Detection of floating plastics from satellite and unmanned aerial systems (Plastic Litter Project 2018)." *Int J Appl Earth Obs Geoinf* 79:175-183.

Topouzelis, K., D. Papageorgiou, A. Karagaitanakis, A. Papakonstantinou, and M. A. Ballesteros. 2020. "Remote sensing of sea surface artificial floating plastic targets with Sentinel-2 and unmanned aerial systems (Plastic Litter Project 2019)." *Remote Sens* 12 (12). doi: 10.3390/rs12122013.

Tourinho, P. S., J. A. Ivar do Sul, and G. Fillmann. 2010. "Is marine debris ingestion still a problem for the coastal marine biota of southern Brazil?" *Mar Pollut Bull* 60 (3):396-401. doi: 10.1016/j.marpolbul.2009.10.013.

Tramoy, R., J. Gasperi, R. Dris, L. Colasse, C. Fisson, S. Sananes, V. Rocher, and B. Tassin. 2019. "Assessment of the plastic inputs from the Seine Basin to the sea using statistical and field approaches." *Front Mar Sci* 6. doi: 10.3389/fmars.2019.00151.

Tunnell, J. W., K. H. Dunning, L. P. Scheef, and K. M. Swanson. 2020. "Measuring plastic pellet (nurdle) abundance on shorelines throughout the Gulf of Mexico using citizen scientists: Establishing a platform for policy-relevant research." *Mar Pollut Bull* 151. doi: 10.1016/j.marpolbul.2019.110794.

Tunnell, K. D. 2008. "Illegal dumping: Large and small scale littering in rural Kentucky." *J Rural Soc Sci* 23 (2):3.

U.S. Census Bureau. 2019. "QuickFacts: California." https://www.census.gov/quickfacts/CA.

U.S. Census Bureau. 2021. "USA Trade Online." https://usatrade.census.gov/.

U.S. Commission on Ocean Policy. 2004. *An Ocean Blueprint for the 21st Century.* Washington, DC.

U.S. Department of Energy. 2021. Plastics Innovation Challenge Draft Roadmap. Washington, D.C.

U.S. EPA. 2002. *Assessing and Monitoring Floatable Debris.* https://www.epa.gov/sites/default/files/2018-12/documents/assess-monitor-floatable-debris.pdf.

U.S. EPA. 2015. U.S. EPA Sustainable Materials Management Program Strategic Plan: Fiscal Year 2017-2022. https://www.epa.gov/sites/default/files/2016-03/documents/smm_strategic_plan_october_2015.pdf.

U.S. EPA. 2020a. Advancing Sustainable Materials Management: 2018 Tables and Figures. https://www.epa.gov/sites/default/files/2021-01/documents/2018_tables_and_figures_dec_2020_fnl_508.pdf.

U.S. EPA. 2020b. "NPDES Stormwater Program." https://www.epa.gov/npdes/npdes-stormwater-program.

U.S. EPA. 2020c. The United States Federal Strategy for Addressing the Global Issue of Marine Litter. https://www.epa.gov/sites/default/files/2020-10/documents/marinelitter_booklet_10.16.20_v10epa.pdf.

U.S. EPA. 2020d. "What are Biobased Plastics?" https://www.epa.gov/trash-free-waters/frequently-asked-questions-about-plastic-recycling-and-composting#biobased.

U.S. EPA. 2021a. "Environmental Justice." https://www.epa.gov/environmentaljustice.

U.S. EPA. 2021b. EPA Administrator Announces Agency Actions to Advance Environmental Justice: Administrator Regan Directs Agency to Take Steps to Better Serve Historically Marginalized Communities. Washington, D.C.

U.S. EPA. 2021c. "Facts and Figures about Materials, Waste and Recycling." https://www.epa.gov/facts-and-figures-about-materials-waste-and-recycling/containers-and-packaging-product-specific-data.

U.S. EPA. 2021d. "International Agreements on Transboundary Shipments of Hazardous Waste." https://www.epa.gov/hwgenerators/international-agreements-transboundary-shipments-hazardous-waste.

U.S. EPA. 2021e. "National Overview: Facts and Figures on Materials, Wastes and Recycling." https://www.epa.gov/facts-and-figures-about-materials-waste-and-recycling/national-overview-facts-and-figures-materials#NationalPicture.

U.S. EPA. 2021f. "Nonpoint Source: Roads Highways and Bridges." https://www.epa.gov/nps/nonpoint-source-roads-highways-and-bridges.

U.S. EPA. 2021g. "Organic Chemicals, Plastics and Synthetic Fibers Effluent Guidelines." https://www.epa.gov/eg/organic-chemicals-plastics-and-synthetic-fibers-effluent-guidelines.

U.S. EPA. 2021h. "Urbanization and Storm Water Runoff." https://www.epa.gov/sourcewaterprotection/urbanization-and-storm-water-runoff.

U.S. EPA. 2021i. "What is the Toxics Release Inventory?" https://www.epa.gov/toxics-release-inventory-tri-program/what-toxics-release-inventory.

U.S. EPA. 2021j "What Types of Plastic Scrap and Waste Are Controlled Under the Basel Convention?" https://www.epa.gov/hwgenerators/new-international-requirements-export-and-import-plastic-recyclables-and-waste#fq3.

U.S. GAO (United States Government Accountability Office). 2019. Marine Debris: Interagency Committee Members Are Taking Action, but Additional Steps Could Enhance the Federal Response.

U.S. GAO. 2020. *Recycling: Building on Existing Federal Efforts Could Help Address Cross-Cutting Challenges.* GAO-21-87.

Uhrin, A. V., S. Lippiatt, C. E. Herring, K. Dettloff, K. Bimrose, and C. Butler-Minor. 2020. "Temporal trends and potential drivers of stranded marine debris on beaches within two US National Marine Sanctuaries using citizen science data." *Front Environ Sci* 8. doi: 10.3389/fenvs.2020.604927.

UNEP (United Nations Environment Programme). 2014. *Valuing Plastics: The Business Case for Measuring, Managing and Disclosing Plastic Use in the Consumer Goods Industry.*

UNEP. 2018. *Legal Limits on Single-Use Plastics and Microplastics: A Global Review of National Laws and Regulations,* edited by C. Excell, C. Salcedo-La Viña, J. Worker, and E. Moses. United Nations Environment Programme and World Resources Institute.

UNEP. 2020. *National Guidance for Plastic Pollution Hotspotting and Shaping Action: Introduction to the Methodology.* Nairobi, Kenya: United Nations Environment Programme.

UNEP. 2021a. *From Pollution to Solution: A Global Assessment of Marine Litter and Plastic Pollution.* Nairobi, Kenya: United Nations Environment Programme.

UNEP. 2021b. *Neglected Environmental Justice Impacts of Marine Litter and Plastic Pollution,* edited by J. Calil, M. Gutiérrez-Graudiņš, S. Munguía, and C. Chin. Nairobi, Kenya: United Nations Environmental Programme.

Unger, B., E. L. B. Rebolledo, R. Deaville, A. Gröne, L. L. IJsseldijk, M. F. Leopold, U. Siebert, J. Spitz, P. Wohlsein, and H. Herr. 2016. "Large amounts of marine debris found in sperm whales stranded along the North Sea coast in early 2016." *Mar Pollut Bull* 112 (1-2):134-141. doi: 10.1016/j.marpolbul.2016.08.027.

United Church of Christ. 1987. *Toxic Wastes and Race in the United States: A National Report on the Racial and Socio-economic Characteristics of Communities with Hazardous Waste Sites.* https://www.nrc.gov/docs/ML1310/ML13109A339.pdf.

Upadhyaya, N. 2019. "Recycling Is Going to Waste!". Atlantic Council. https://www.atlanticcouncil.org/blogs/energysource/recycling-is-going-to-waste/.

van Cauwenberghe, L., and C. R. Janssen. 2014. "Microplastics in bivalves cultured for human consumption." *Environ Pollut* 193:65-70. doi: 10.1016/j.envpol.2014.06.010.

van Cauwenberghe, L., A. Vanreusel, J. Mees, and C. R. Janssen. 2013. "Microplastic pollution in deep-sea sediments." *Environ Pollut* 182:495-499.

van der Velde, T., D. A. Milton, T. J. Lawson, C. Wilcox, M. Lansdell, G. Davis, G. Perkins, and B. D. Hardesty. 2017. "Comparison of marine debris data collected by researchers and citizen scientists: Is citizen science data worth the effort?" *Biol Conserv* 208:127-138. doi: 10.1016/j.biocon.2016.05.025.

van Emmerik, T., T. C. Kieu-Le, M. Loozen, K. Oeveren, E. Strady, X. T. Bui, M. Egger, J. Gasperi, L. Lebreton, P. D. Nguyen, A. Schwarz, B. Slat, and B. Tassin. 2018. "A methodology to characterize riverine macroplastic emission into the ocean." *Front Mar Sci.* doi: 10.3389/fmars.2018.00372.

van Franeker, J. A., and K. L. Law. 2015. "Seabirds, gyres and global trends in plastic pollution." *Environ Pollut* 203:89-96. doi: 10.1016/j.envpol.2015.02.034.

van Sebille, E., C. Wilcox, L. Lebreton, N. Maximenko, B. D. Hardesty, J. A. van Franeker, M. Eriksen, D. Siegel, F. Galgani, and K. L. Law. 2015. "A global inventory of small floating plastic debris." *Environ Res Lett* 10 (12). doi: 10.1088/1748-9326/10/12/124006.

van Sebille, E., S. Aliani, K. L. Law, N. Maximenko, J. M. Alsina, A. Bagaev, M. Bergmann, B. Chapron, I. Chubarenko, A. Cózar, P. Delandmeter, M. Egger, B. Fox-Kemper, S. P. Garaba, L. Goddijn-Murphy, B. D. Hardesty, M. J. Hoffman, A. Isobe, C. E. Jongedijk, M. L. A. Kaandorp, L. Khatmullina, A. A. Koelmans, T. Kukulka, C. Laufkötter, L. Lebreton, D. Lobelle, C. Maes, V. Martinez-Vicente, M. A. Morales Maqueda, M. Poulain-Zarcos, E. Rodríguez, P. G. Ryan, A. L. Shanks, W. J. Shim, G. Suaria, M. Thiel, T. S. van den Bremer, and D. Wichmann. 2020. "The physical oceanography of the transport of floating marine debris." *Environ Res Lett* 15 (2). doi: 10.1088/1748-9326/ab6d7d.

Vegter, A. C., M. Barletta, C. Beck, J. Borrero, H. Burton, M. L. Campbell, M. F. Costa, M. Eriksen, C. Eriksson, A. Estrades, K. V. K. Gilardi, B. D. Hardesty, J. A. Ivar do Sul, J. L. Lavers, B. Lazar, L. Lebreton, W. J. Nichols, C. A. Ribic, P. G. Ryan, Q. A. Schuyler, S. D. A. Smith, H. Takada, K. A. Townsend, C. C. C. Wabnitz, C. Wilcox, L. C. Young, and M. Hamann. 2014. "Global research priorities to mitigate plastic pollution impacts on marine wildlife." *Endanger Species Res* 25 (3):225-247. doi: 10.3354/esr00623.

Venrick, E. L., T. W. Backman, W. C. Bartram, C. J. Platt, M. S. Thornhill, and R. E. Yates. 1973. "Man-made objects on the surface of the Central North Pacific Ocean." *Nature* 241 (5387):271-271. doi: 10.1038/241271a0.

Verma, R., K. S. Vinoda, M. Papireddy, and A. N. S. Gowda. 2016. "Toxic pollutants from plastic waste- A review." *Procedia Environ Sci* 35:701-708. doi: https://doi.org/10.1016/j.proenv.2016.07.069.

Villarrubia-Gómez, P., S. E. Cornell, and J. Fabres. 2018. "Marine plastic pollution as a planetary boundary threat – The drifting piece in the sustainability puzzle." *Mar Policy* 96:213-220. doi: https://doi.org/10.1016/j.marpol.2017.11.035.

Wagner, S., P. Klockner, B. Stier, M. Romer, B. Seiwert, T. Reemtsma, and C. Schmidt. 2019. "Relationship between discharge and river plastic concentrations in a rural and an urban catchment." *Environ Sci Technol* 53 (17):10082-10091. doi: 10.1021/acs.est.9b03048.

Walls, M., and K. Palmer. 2001. "Upstream pollution, downstream waste disposal, and the design of comprehensive environmental policies." *J Environ Econ and Management* 41 (1):94-108. doi: 10.1006/jeem.2000.1135.

Wang, S., K. A. Lydon, E. M. White, J. B. Grubbs, 3rd, E. K. Lipp, J. Locklin, and J. R. Jambeck. 2018. "Biodegradation of poly(3-hydroxybutyrate-co-3-hydroxyhexanoate) plastic under anaerobic sludge and aerobic seawater conditions: Gas evolution and microbial diversity." *Environ Sci Technol* 52 (10):5700-5709. doi: 10.1021/acs.est.7b06688.

Wang, W., and J. Wang. 2018. "Investigation of microplastics in aquatic environments: An overview of the methods used, from field sampling to laboratory analysis." *TrAC Trends Anal Chem* 108:195-202. doi: 10.1016/j.trac.2018.08.026.

Waste Advantage Magazine. 2020. "The Impact of COVID-19 on the Waste and Recycling Industry." May 1.

Waste Atlas (2016) D-waste. http://www.atlas.d-waste.com.

Watkins, L., S. McGrattan, P. J. Sullivan, and M. T. Walter. 2019. "The effect of dams on river transport of microplastic pollution." *Sci Total Environ* 664:834-840. doi: 10.1016/j.scitotenv.2019.02.028.

Watson, S. 2018. "Leo Hendrik Baekland." Encyclopedia.com. https://www.encyclopedia.com/science/encyclopedias-almanacs-transcripts-and-maps/leo-hendrik-baekeland.

Werbowski, L. M., A. N. Gilbreath, K. Munno, X. Zhu, J. Grbic, T. Wu, R. Sutton, M. D. Sedlak, A. D. Deshpande, and C. M. Rochman. 2021. "Urban stormwater runoff: A major pathway for anthropogenic particles, black rubbery fragments, and other types of microplastics to urban receiving waters." *ACS ES&T Water* 1 (6):1420-1428. doi: 10.1021/acsestwater.1c00017.

Wessel, C. C., G. R. Lockridge, D. Battiste, and J. Cebrian. 2016. "Abundance and charac-teristics of microplastics in beach sediments: Insights into microplastic accumulation in northern Gulf of Mexico estuaries." *Mar Pollut Bull* 109 (1):178-183. doi: 10.1016/j.marpolbul.2016.06.002.

White, R. 2018. *Life at the Fenceline: Understanding Cumulative Health Hazards in Environmental Justice Communities.* Coming Clean, Environmental Justice Health Alliance for Chemical Policy Reform, and Campaign for Healthier Solutions.

Whitney, J. L., J. M. Gove, M. A. McManus, K. A. Smith, J. Lecky, P. Neubauer, J. E. Phipps, E. A. Contreras, D. R. Kobayashi, and G. P. Asner. 2021. "Surface slicks are pelagic nurseries for diverse ocean fauna." *Sci Rep* 11 (1):3197. doi: 10.1038/s41598-021-81407-0.

Wiesinger, H., Z. Wang, and S. Hellweg. 2021. "Deep dive into plastic monomers, additives, and processing aids." *Environ Sci Technol* 55 (13):9339-9351. doi: 10.1021/acs.est.1c00976.

Williams, A. T., and S. L. Simmons. 1997. "Estuarine litter at the river/beach interface in the Bristol Channel, United Kingdom." *J Coast Res* 13 (4):1159-1165.

Williams, A. T., S. L. Simmons, and A. Fricker. 1993. "Off-shore sinks of marine litter: A new problem." *Mar Pollut Bull* 26 (7):404-405. doi: 10.1016/0025-326x(93)90192-m.

Windsor, F. M., R. M. Tilley, C. R. Tyler, and S. J. Ormerod. 2019. "Microplastic inges-tion by riverine macroinvertebrates." *Sci Total Environ* 646:68-74. doi: 10.1016/j.scitotenv.2018.07.271.

Wolff, S., Jutta Kerpen, J. Prediger, L. Barkmann, and L. Müller. 2019. "Determination of the microplastics emission in the effluent of a municipal waste water treatment plant using Raman microspectroscopy." *Water Res X* 2:100014.

Woodall, L. C., L. F. Robinson, A. D. Rogers, B. E. Narayanaswamy, and G. L. Paterson. 2015. "Deep-sea litter: A comparison of seamounts, banks and a ridge in the Atlantic and Indian Oceans reveals both environmental and anthropogenic factors impact accumu-lation and composition." *Front Mar Sci* 2. doi: 10.3389/fmars.2015.00003.

Woods, J. S., G. Rødder, and F. Verones. 2019. "An effect factor approach for quantifying the entanglement impact on marine species of macroplastic debris within life cycle impact assessment." *Ecol Indic* 99:61-66. doi: 10.1016/j.ecolind.2018.12.018.

Woodward, J., J. Li, J. Rothwell, and R. Hurley. 2021. "Acute riverine microplastic contamination due to avoidable releases of untreated wastewater." *Nat Sustain.* doi: 10.1038/s41893-021-00718-2.

World Bank. 2021. "Population, total." https://data.worldbank.org/indicator/SP.POP.TOTL?most_recent_value_desc=true.

World Economic Forum, Ellen MacArthur Foundation, and McKinsey & Company. 2016. "The New Plastics Economy — Rethinking the Future of Plastics." https://ellenmacarthurfoundation.org/the-new-plastics-economy-rethinking-the-future-of-plastics.

Wright, R. J., G. Erni-Cassola, V. Zadjelovic, M. Latva, and J. A. Christie-Oleza. 2020. "Marine plastic debris: A new surface for microbial colonization." *Environ Sci Technol* 54 (19):11657-11672. doi: 10.1021/acs.est.0c02305.

Wright, S. L., R. C. Thompson, and T. S. Galloway. 2013. "The physical impacts of micro-plastics on marine organisms: A review." *Environ Pollut* 178:483-492. doi: 10.1016/j.envpol.2013.02.031.

Wyer, H. 2021. Action Item: Discussion and Possible Endorsement of Recommendations to Address Plastic Pollution in California's Coastal and Marine Ecosystems. Ocean Protection Council.

Yang, R., H. Wei, J. Guo, and A. Li. 2012. "Emerging brominated flame retardants in the sed-iment of the Great Lakes." *Environ Sci Technol* 46 (6):3119-3126. doi: 10.1021/es204141p.

Yonkos, L. T., E. A. Friedel, A. C. Perez-Reyes, S. Ghosal, and C. D. Arthur. 2014. "Micro-plastics in four estuarine rivers in the Chesapeake Bay, U.S.A." *Environ Sci Technol* 48 (24):14195-14202. doi: 10.1021/es5036317.

Yoshida, S., K. Hiraga, T. Takehana, I. Taniguchi, H. Yamaji, Y. Maeda, K. Toyohara, K. Miyamoto, Y. Kimura, and K. Oda. 2016. "A bacterium that degrades and assimilates poly(ethylene terephthalate)." *Science* 351 (6278):1196-1199. doi: 10.1126/science. aad6359.

Youngblood, K., S. Finder, and J Jambeck. 2021. *Mississippi River Plastic Pollution Initiative 2021 Science Report.* Athens, GA: University of Georgia.

Youngblood, K. M., A. L. Brooks, N. Das, A. Singh, M. Sultana, G. Verma, T. Zakir, G. Chowdhury, E. Duncan, H. Khatoon, T. Maddalene, I. Napper, S. Nelms, S. Patel, V. Sturges, and J. Jambeck. In Review. "Rapid characterization of land-based litter: Leakage and input in the Ganges River basin." *Environ Sci Technol.*

Yu, J. S., B. S. Yoon, J. H. Rho, S. H. Yoon, and Energy. 2002. "Investigation of floating debris characteristics drained from 4 big rivers on a flooding." *J Korean Soc Mar Environ* 5 (3):45-53.

Zarfl, C. 2019. "Promising techniques and open challenges for microplastic identification and quantification in environmental matrices." *Anal Bioanal Chem* 411 (17):3743-3756.

Zettler, E. R., T. J. Mincer, and L. A. Amaral-Zettler. 2013. "Life in the "plastisphere": Microbial communities on plastic marine debris." *Environ Sci Techno* 47 (13):7137-7146. doi: 10.1021/es401288x.

Zhang, K., X. Xiong, H. Hu, C. Wu, Y. Bi, Y. Wu, B. Zhou, P. K. Lam, and J. Liu. 2017. "Occurrence and characteristics of microplastic pollution in Xiangxi Bay of Three Gorges Reservoir, China." *Environ Sci Technol* 51 (7):3794-3801. doi: 10.1021/acs. est.7b00369.

Zhao, S., E. R. Zettler, L. A. Amaral-Zettler, and T. J. Mincer. 2021. "Microbial carrying capacity and carbon biomass of plastic marine debris." *ISME J* 15 (1):67-77. doi: 10.1038/ s41396-020-00756-2.

Zheng, J., and S. Suh. 2019. "Strategies to reduce the global carbon footprint of plastics." *Nat Cli Chang* 9 (5):374-378. doi: 10.1038/s41558-019-0459-z.

Zheng, Y., J. Li, W. Cao, X. Liu, F. Jiang, J. Ding, X. Yin, and C. Sun. 2019. "Distribution characteristics of microplastics in the seawater and sediment: A case study in Jiaozhou Bay, China." *Sci Total Environ* 674:27-35. doi: 10.1016/j.scitotenv.2019.04.008.

Zimmerman, J. B., P. T. Anastas, H. C. Erythropel, and W. Leitner. 2020. "Designing for a green chemistry future." *Science* 367 (6476):397-400. doi: 10.1126/science.aay3060.

Appendixes

A

Biographies of the Committee on the United States Contributions to Global Ocean Plastic Waste

Margaret Spring
Chair

Margaret Spring is chief conservation and science officer at Monterey Bay Aquarium, with decades of experience in environmental law and policy. She oversees the Monterey Bay Aquarium's conservation, science, and markets programs, including Seafood Watch, and coordinates the aquarium's environmental sustainability initiatives. Before joining the aquarium in 2013, Spring served as chief of staff, and later principal deputy under secretary, at the U.S. National Oceanic and Atmospheric Administration (NOAA) from 2009 to 2013. Prior to her tenure at NOAA, Spring led the Nature Conservancy's California coastal and marine program from 2007 to 2009. Spring served for 8 years (1999–2007) as senior and general counsel to the U.S. Senate Committee on Commerce, Science, and Transportation where she advised members of Congress and developed key ocean and climate legislation, including the Magnuson-Stevens Fishery Conservation and Management Reauthorization Act of 2006 and the Oceans Act of 2000, which created the U.S. Commission on Ocean Policy. From 1992 to 1999, Spring was an environmental attorney in private practice at Sidley & Austin in Washington, D.C., specializing in clean water and hazardous waste matters. She currently serves on the boards of the Environmental Law Institute, the California Ocean Science Trust, and the Monterey Bay Fisheries Trust. She previously served as a member of the Ocean Studies Board (2014–2020) and chaired the American Geophysical Union's Position Statement Committee from 2015 to 2019.

She is a graduate of Duke University Law School and Dartmouth College and was a 1991 John A. Knauss Sea Grant Fellow in the U.S. Senate.

Mary J. Donohue
Member

Dr. Mary Donohue serves as specialist faculty at the University of Hawai'i Sea Grant College Program where she conducts research, extension, communications, and program and project management. She also serves as affiliate faculty at the Environmental Sciences Graduate Program at Oregon State University. She is trained as a marine mammal physiological ecologist with interests in marine mammal conservation, including the effects of derelict fishing gear and microplastics on marine mammals and the environment. Dr. Donohue has also focused on activities toward achieving sustainable communities through understanding and mitigating pollution in water resources. Her graduate research was conducted on the Pribilof Islands, Alaska where she investigated the physiological and behavioral ontogeny of northern fur seals. On the Pribilof Islands she observed the problem of plastic pollution on remote oceanic islands, including the entanglement of seals, foxes, and birds in plastic waste and derelict fishing gear. This experience informed her later position with the U.S. National Oceanic and Atmospheric Administration where from 1998 to 2002 she developed, administered, coordinated, and served as chief scientist on the first systematic at-sea expeditions to document, study, and remove large aggregations of derelict fishing gear and other marine debris from the coral reefs of the Northwestern Hawaiian Islands, habitat of the critically endangered Hawaiian monk seal, threatened and endangered marine turtles, and other wildlife. Dr. Donohue had the privilege of experiencing an at-sea arrested landing and catapult assisted takeoff by air on the USS *Abraham Lincoln* (CVN 72) Nimitz-class aircraft carrier as an environmental expert to observe the U.S. Navy's implementation of a shipboard plastic waste management system. Dr. Donohue has published broadly on marine mammals, marine debris, water quality, and, more recently, workforce development, with an emphasis on science, technology, engineering, and mathematics participation, including that relating to justice, equity, diversity, and inclusion. She has most recently served as senior author on a strategic visioning document for the United States Geological Survey Water Resources Research Act Program, which will guide and direct national, regional, and state activities of the program for the next 10 years. She has spoken at international conferences and symposia and as an invited university and public seminar speaker in the United States, Japan, Canada, Spain, Monaco, Malta, and Scotland. Dr. Donohue previously served as a committee member for the National Academies Committee on the Effectiveness of International and National Measures to Reduce and Prevent Marine Debris and Its Impacts in 2007–2008. She began her academic training at Santa Barbara City College, later earning a B.A. degree in aquatic biology from the University of California, Santa Barbara in 1989 and M.A. and Ph.D. degrees from the University of California, Santa Cruz in organismal and population biology in 1997 and 1998, respectively.

Michelle Gierach
Member

Dr. Michelle Gierach is a senior scientist in the Water and Ecosystems Group at the National Aeronautics and Space Administration (NASA) Jet Propulsion Laboratory. Her research interests include analysis and application of multispectral and hyperspectral (otherwise referred to as imaging spectroscopy) airborne and spaceborne observations to study synoptic to decadal changes in the aquatic environment. She has been involved in several NASA satellite and airborne remote sensing missions, including but not limited to, as co-lead for the NASA Surface Biology and Geology mission Pathfinder study (SISTER), project scientist for the NASA Earth Venture Suborbital 2 (EVS-2) Coral Reef Airborne Laboratory mission, and science team member for the NASA EVS-3 Sub-Mesoscale Ocean Dynamics Experiment mission. She is currently a member of the International Ocean Color Coordinating Group Task Force: Remote Sensing of Marine Litter and Debris, and co-chair of the U.S. CLIVAR Phenomena, Observations, and Synthesis Panel. She earned a B.S. and an M.S. in meteorology from Florida State University in 2004 and 2006, respectively, and a Ph.D. in marine science from the University of South Carolina in 2009.

Jenna Jambeck
Member

Dr. Jenna Jambeck is a Georgia Athletic Association Distinguished Professor in Environmental Engineering in the College of Engineering at the University of Georgia (UGA), Lead of the Center for Circular Materials Management and Circularity Informatics Lab in the New Materials Institute at UGA and a National Geographic Fellow. She has been conducting research on solid waste issues for more than 24 years with related projects on marine debris since 2001. She also specializes in global waste management issues and plastic contamination. Her work on plastic waste inputs into the ocean has been recognized by the global community and translated into policy discussions by the Global Ocean Commission, in testimony to U.S. Congress, in G7 and G20 Declarations, and the United Nations Environment program. She conducts public environmental diplomacy as an international informational speaker for the

U.S. Department of State. This has included multiple global programs of speaking events, meetings, presentations to governmental bodies, and media outreach in 13 countries. She has won awards for her teaching and research in the College of Engineering and the UGA Creative Research Medal, as well as a Public Service and Outreach Fellowship. She received her master's and doctorate degrees in environmental engineering from the University of Florida in 1998 and 2004, respectively. She graduated with bachelor's degree in environmental engineering with honors from Florida in 1996.

Hauke Kite-Powell
Member

Dr. Hauke L. Kite-Powell is a research specialist at the Marine Policy Center of the Woods Hole Oceanographic Institution. Dr. Kite-Powell also holds appointments as a lecturer at the Massachusetts Maritime Academy and as a senior analyst with Marsoft Inc. Dr. Kite-Powell's research focuses on public- and private-sector management issues for marine resources and the economic activities that depend on them. Current and recent research projects include work on costs and benefits from improved ocean observing activities; approaches to economic valuation of marine resources; policy issues surrounding use of ocean "space" for non-traditional activities, such as aquaculture and wind power; economics and management of marine aquaculture operations; economics of ocean plastics and removal of plastics from the oceans; and economic dimensions of climate change effects on marine ecosystems, shoreline change, and the carbon cycle. Dr. Kite-Powell has contributed to several National Academies studies including Charting a Course into the Digital Era: Guidance for NOAA's Nautical Charting Mission (1994); Critical Infrastructure for Ocean Research and Societal Needs in 2030 (2011); Best Practices for Shellfish Mariculture and the Effects of Commercial Mariculture on Drakes Estero, Pt. Reyes National Seashore, California (2010); and Ecosystem Concepts for Sustainable Bivalve Mariculture (2010). He holds degrees in naval architecture (B.S), technology and policy (M.S.), and ocean systems management (M.S. and Ph.D.) from the Massachusetts Institute of Technology.

Kara Lavender Law
Member

Dr. Kara Lavender Law is Research Professor of Oceanography at Sea Education Association (SEA; Woods Hole, Massachusetts). Since 2007, Dr. Law's research has focused on plastic debris in the ocean, beginning with an analysis of SEA's then-25-year data set of floating microplastics in the

North Atlantic, collected by more than 7,000 SEA students and scientists. Her initial research focused on physical processes that transport and transform plastics in the marine environment, and has since expanded "upstream" to better understand the generation, pathways, and treatment of plastic waste, with the goal to ultimately prevent plastics from leaking into the environment. Dr. Law served as co-principal investigator of the Marine Debris Working Group at the National Center for Ecological Analysis and Synthesis, is co-chair of the SCOR Working Group FLOTSAM (Floating Litter and its Oceanic TranSport Analysis and Modelling), and has participated in many other international working groups, workshops, and panels, including at the National Academies, on the topic of plastic marine debris. Dr. Law holds several scientific advisory roles and strives to effectively communicate the scientific understanding of ocean plastics, including major knowledge gaps, to wide-ranging audiences including policy makers, industry groups, students, and the general public. In 2018 she served as a witness in the U.S. Senate Committee on Environment and Public Works Hearing on "Cleaning Up the Oceans: How to Reduce the Impact of Man-Made Trash on the Environment, Wildlife, and Human Health?" Dr. Law received her Ph.D. from the Scripps Institution of Oceanography/University of California, San Diego in physical oceanography (2001), and a B.S. in mathematics from Duke University (1994).

Jay R. Lund
Member

Dr. Jay R. Lund (NAE) is co-director of the Center for Watershed Sciences and Distinguished Professor of Civil and Environmental Engineering, Department of Civil and Environmental Engineering, at the University of California (UC), Davis. Dr. Lund joined the UC Davis faculty in 1987. He teaches and conducts research on applications of systems analysis, economic, and management methods to infrastructure and public works problems. His recent work is primarily on water and environmental problems, but he has done substantial work in solid and hazardous waste management; dredging and coastal zone management; and urban, regional, and transportation planning. While most of this work involves the application of economics, optimization, and simulation modeling, his interests also include more qualitative policy, planning, and management studies. He was elected to the National Academy of Engineering for analysis of water and environmental policy issues leading to integrated water resources planning and management. He served on the Committee on Further Studies of Endangered and Threatened Fishes in the Klamath River and the Committee to Review the New York City Watershed Protection Program. Dr. Lund has a B.A. in regional planning and international relations from the University of Delaware (1979). He also has a B.S. in civil engineering, an M.A. in geography (1983), and a Ph.D. in civil engineering, all from the University of Washington (1986).

Ramani Narayan
Member

Dr. Ramani Narayan is University Distinguished Professor at Michigan State University (MSU) in the Department of Chemical Engineering and Materials Science. He has more than 200 refereed publications in leading journals and 32 issued patents, and he edited three books in the area of environmentally responsible biobased materials (h-index 50, i10-index 143, 12,248 citations in Google Scholar). He has graduated 22 Ph.D. and 23 master's students at MSU and currently has 4 Ph.D. students working in his group along with several postdoctoral fellows, industrial visiting fellows, and 6 undergraduate students. He has won many awards and honors including fellow of U.S. National Academy of Inventors, the MSU University Distinguished Professor in 2007, and the William N. Findley Award for "significant contributions to the application of new technologies within the scope of ASTM Committee D20 on Plastic." Dr. Narayan received his master's in organic chemistry and Ph.D. in organic chemistry (polymer science and engineering) from Bombay University.

Eben Schwartz
Member

Eben Schwartz, a staff member of the California Coastal Commission since 2000, runs Marine Debris and Public Outreach programs for the Public Education Program. Schwartz has a lengthy history of work on plastic pollution and marine debris, topics on which he has become one of the state's leading authorities. As the leader of California Coastal Cleanup Day, California's largest volunteer event, as well as the year-round Adopt-a-Beach Program, Schwartz has helped motivate hundreds of thousands of citizens to become active in the fight against plastic pollution. Schwartz works with hundreds of nonprofit organizations, local government agencies, and corporate partners to coordinate beach and inland shoreline cleanups as well as develop long-term policies that will stop pollution at its source. Schwartz serves as the chair of the West Coast Marine Debris Alliance, an organization that he helped found in 2008, initially as part of the West Coast Governors' Alliance on Ocean Health. From 2007 to 2010, Schwartz served as the chair of the California Ocean Protection Council's Marine Debris Steering Committee while it was drafting and adopting the first statewide Ocean Litter Strategy, and currently serves on the planning committee for that strategy's update, which was adopted and published in April 2018. Schwartz has traveled extensively to give talks and presentations about the challenges and potential solutions to marine debris and plastic pollution. A brief sampling includes an address to *The Economist's* Sustainability Summit in London in March 2019, a keynote address to the United Nations Environment Programme's Northwest Pacific Action Plan Marine Litter Workshop in Okinawa, Japan, in October 2013, and multiple presentations to both the 5th and 6th Marine Debris Conferences in 2011 and 2018. In 2018, the U.S. State Department's Bureau of Educational and Cultural Affairs invited Schwartz to join its Guest Speaker Program, for which he has traveled to Poland in 2019 and participated in virtual programs in Taiwan in fall 2020. Prior to his work with the Coastal Commission, Schwartz worked in conservation programs at the Sierra Club at both the local and national levels. In 2007, Schwartz was awarded an Aspen Institute Fellowship and served as one of the inaugural Catto Fellows, a program designed for emerging leaders in the environment and energy sector. Schwartz holds a B.A. from Johns Hopkins University.

Rashid Sumaila
Member

Dr. Rashid Sumaila is a professor and Canada Research Chair (Tier 1) in Interdisciplinary Ocean and Fisheries Economics. He is director of both the Fisheries Economics Research Unit and the OceanCanada Partnership at the Institute for the Oceans and Fisheries, University of British Columbia (UBC). Dr. Sumaila is also appointed in the UBC School of Public Policy and Global Affairs. His research focuses on bioeconomics, marine ecosystem valuation, and the analysis of global issues such as fisheries subsidies, illegal fishing, climate change, marine plastic pollution, and oil spills. Dr. Sumaila is widely published and cited. He is on the editorial boards of several journals, including *Science Advances*, *Scientific Reports*, and *Environmental & Resource Economics*. As well as winning the 2017 Volvo Environment Prize and other prestigious awards, Dr. Sumaila was inducted into the Fellowship of the Royal Society of Canada in 2019. He was named a Hokkaido University Ambassador in 2016 and a Distinguished Professor (visiting) at the National University of Malaysia in 2020. Dr. Sumaila has given talks at the UN Rio+20, the World Trade Organization, the White House, the Canadian Parliament, the African Union, the St. James Palace, and the British House of Lords. Dr. Sumaila has served on many scientific advisory boards and high-level panels. He is currently on the board of directors of Oceana and he is a member of the science advisory committee for the High-Level Panel for a Sustainable Ocean Economy convened by the Prime Minister of Norway, which consists of 14 sitting heads of states and governments. The mission of the panel is to build momentum toward a sustainable ocean economy, where "effective protection, sustainable production and equitable prosperity go hand-in-hand." Dr. Sumaila received his Ph.D. and M.Sc. from the University of Bergen, Norway, in economics in 1996 and 1993, respectively, and graduated with a B.Sc. (Hon.) degree in quantity surveying from the Ahmadu Bello University, Nigeria, in 1986.

B

Definitions and Acronyms

Leakage: Loss of custodial control of plastic material to the environment, including during routine activities.

Marine debris or marine litter: Any persistent, manufactured, or processed solid material that is directly or indirectly, intentionally or unintentionally, discarded, disposed of, or abandoned into the marine, coastal, or Great Lakes environment. This definition excludes natural flotsam, such as trees washed out to sea, and focuses on non-biodegradable synthetic materials that persist in the marine environment (definition adapted from multiple sources).

Microplastic: A plastic object from 1 to 1,000 µm in size as determined by the object's largest dimension (definition adapted from Hartmann et al. 2019).

Ocean plastic waste: A subset of marine debris; plastic waste in the marine environment including estuaries, coastlines, seawater (sea surface and water column), seafloor sediments, biota, and sea ice (these are similar ocean reservoirs as defined in Law 2017).

Ocean plastic waste, plastic marine debris, plastic marine litter, and marine plastic pollution are collapsed for clarity and used interchangeably.

Plastic solid waste: The subset of solid waste that is composed of plastics.

Plastic waste: Any plastic that has been intentionally or unintentionally taken out of use and that has entered a waste stream as part of a waste management process or released into the environment. Plastic waste in the environment is typically characterized according to size. Size classifications in this report follow the classifications used by the Joint Group of Experts on the Scientific Aspects of the Marine Environmental Protection (GESAMP 2019) and adopted by the National Oceanic and Atmospheric Administration Marine Debris Program.

Plastics: A wide range of synthetic polymeric materials and associated additives made from petrochemical, natural gas, or biologically based feedstocks and with thermoplastic, thermoset, or elastomeric properties used in a wide variety of applications including packaging, building and construction, household and sports equipment, vehicles, electronics, and agriculture, and which occur in a solid state in the environment.

Solid waste: Residential, commercial, and institutional waste (Kaza et al. 2018). Industrial, medical, hazardous, electronic, and construction and demolition waste are excluded from this definition.

Virgin plastic: Plastic resin produced from a petrochemical, natural gas, or biobased feedstock, which has never been used or processed.

ACC	American Chemistry Council
ALDFG	abandoned, lost, or otherwise discarded fishing gear
ASTM	ASTM International (formerly American Society for Testing and Materials)
BMT	billion metric tons
CERCLA	Comprehensive Environmental Response, Compensation and Liability Act
CFCs	chlorofluorocarbons
CFR	Code of Federal Regulations
CWA	Clean Water Act
DDT	dichlorodiphenyltrichloroethane
EPR	extended producer responsibility
EPS	expanded polystyrene
EU	European Union
FTIR	Fourier transform infrared
GAO	Government Accountability Office
GESAMP	Joint Group of Experts on the Scientific Aspects of the Marine Environmental Protection
HBCDs	hexabromocyclododecanes

HDPE	high-density polyethylene
ICC	International Coastal Cleanup
IMDCC	Interagency Marine Debris Coordinating Committee
ISO	International Standards Organization
LDPE	low-density polyethylene
LIDAR	Light Detection and Ranging
LLDPE	linear low-density polyethylene
MDMAP	Marine Debris Monitoring and Assessment Project
MDP	Marine Debris Program
MEE	Ministry of Ecology and Environment
MMT	million metric tons
MRF	material recovery facility
MSW	municipal solid waste
NDPB	non-degradable plastic bags
NDRC	National Development and Reform Commission
NIR	near-infrared
NOAA	National Oceanic and Atmospheric Administration
NSPT	non-degradable single-use plastic tableware
OECD	Organisation for Economic Co-operation and Development
OSB	Ocean Studies Board
PE	polyethylene
PET	polyethylene terephthalate
PP	polypropylene
PPE	personal protective equipment
PS	polystyrene
PVC	polyvinyl chloride
py-GC-MS	pyrolysis-gas chromatography-mass spectrometry
RCRA	Resource Conservation and Recovery Act
RGB	red-green-blue
SOT	statement of task
SWIR	shortwave infrared
TED-GC-MS	thermal extraction-desorption gas chromatography-mass spectrometry
TMDL	Total Maximum Daily Load
TPU	thermoplastic polyurethane
TRI	Toxics Release Inventory
U.S. EPA	United States Environmental Protection Agency
UAV	unmanned aerial vehicle
UNEP	United Nations Environment Programme
USGS	U.S. Geological Survey
UV	ultraviolet

REFERENCES

GESAMP (Joint Group of Experts on the Scientific Aspects of the Marine Environmental Protection). 2019. *Guidelines or the Monitoring and Assessment of Plastic Litter and Microplastics in the Ocean*, edited by P. J. Kershaw, A. Turra, and F. Galgani.

Hartmann, N. B., T. Hüffer, R. C. Thompson, M. Hassellöv, A. Verschoor, A. E. Daugaard, S. Rist, T. Karlsson, N. Brennholt, M. Cole, M. P. Herrling, M. C. Hess, N. P. Ivleva, A. L. Lusher, and M. Wagner. 2019. "Are we speaking the same language? Recommendations for a definition and categorization framework for plastic debris." *Environ Sci Technol* 53 (3):1039-1047. doi: 10.1021/acs.est.8b05297.

Kaza, S., L. Yao, P. Bhada-Tata, and F. Van Woerden. 2018. *What a Waste 2.0: A Global Snapshot of Solid Waste Management to 2050*. Washington, D.C.: World Bank.

Law, K. L. 2017. "Plastics in the marine environment." *Ann Rev Mar Sci* 9:205-229. doi: 10.1146/annurev-marine-010816-060409.

C

Legal Framework[1]

U.S. FEDERAL LAW: APPLICABILITY TO PLASTICS, PLASTIC POLLUTION, OCEAN PLASTIC WASTE, AND MARINE DEBRIS

Starting in the 1970s, the United States created several legal frameworks designed to control and prevent the release of harmful, toxic, or hazardous substances, as well as manage their transportation, treatment, and disposal. Federal law regulates waste disposal and pollution dispersed across political boundaries (by air and water and soil) with various levels of delegation to states and local authorities. A report issued in late 2020 described a U.S. strategy (2020 Strategy) that included legal authorities and roles of certain federal agencies. In 2021, the United States reported the federal legal framework for marine plastic debris as the Resource Conservation and Recovery Act (RCRA), the Clean Water Act (CWA), the Marine Debris Act as amended in 2018, the Save Our Seas 2.0 Act, the Microbead Free Waters Act of 2015, the Toxic Substances Control Act, and the Rivers and Harbors Appropriations Acts (G20 2021, Ministry of Environment, Japan 2020).

The Solid Waste Disposal Act and RCRA treat plastic waste as a subset of municipal solid waste for disposal in landfills or by incineration. CWA and the Clean Air Act address water and air pollution but do not specifically include plastic waste as a regulated pollutant. In 2006, the

---

[1]The prepublication of this report did not include all of the citations used in the preparation of the Appendix C table. The table has been edited to provide more accurate and complete citations.

Marine Debris Research, Prevention, and Reduction Act (Marine Debris Act) was signed into law. The Coast Guard and Maritime Transportation Act of 2012 reauthorized the Marine Debris Act. Save Our Seas Act of 2018 amended and reauthorized the Marine Debris Act, and the Save Our Seas 2.0 Act amended it in 2020. The Marine Debris Act is the most comprehensive legislation in force relating to ocean plastic waste and other marine debris. These laws focus on cleanup, government coordination, outreach, grant making, and research but do not provide specific authority for any federal agency to regulate the production, transportation, or release of plastic waste.

The most specific legislative action around plastic pollution in aquatic and marine environments was the 2015 Microbead Free Waters Act, which prohibits the manufacturing, packaging, and distribution of rinse-off cosmetics and other products, such as toothpaste, containing plastic microbeads. Other federal laws such as the Ocean Dumping Act support global agreements restricting dumping and pollution from ships and vessels, not land-based sources. International law has been amended to control exports of plastic waste under the Basel Convention on the Control of Transboundary Movements of Hazardous Wastes and Their Disposal, but the United States is not a signatory.

Legal cases testing whether microplastic or macroplastic waste are subject to federal, state, and other legal limits or liability, including under common law, are simultaneously working their way through the courts.

States and local governments also play an increasing role in responding to plastic waste. This work and the subsequent table has been adapted from legal background and information from Mary Ellen Ternes and Scott Fulton (Ternes and Fulton 2020). The following tables and information reflect federal authorities and a summary of state actions on plastic waste and related activities (e.g., research and development and monitoring as of October 2021). References are included at the end for additional details.

U.S. STATE AND LOCAL LEGISLATION

In the absence of federal legislation, state and local governments have taken action to address problems associated with plastic waste "leakage" and litter that is finding its way to the environment. These measures are largely related to single-use plastic items found in cleanups and in waterways. These measures include existing plastic bag laws, product bans, extended producer responsibility, container deposit schemes (bottle bills), and recycling. These local "legislative" laboratories are testing the efficacy of different methods, most focused on single-use plastic bags. As of 2019,

continued

U.S. Law and Key Agencies	Intervention Stage	Key Provisions	Gaps, Roles, and Related Activities
		Plastic-Related Authorities	
Marine Debris Research, Prevention, and Reduction Act of 2006 (as amended in 2012, 2018, and 2020) 33 U.S.C. §§ 1951 et seq. **National Oceanic and Atmospheric Administration** (NOAA) **U.S. Environmental Protection Agency** (U.S. EPA) **U.S. State Department** **National Institute of Standards and Technology** (NIST) **Interagency Marine Debris Coordinating Committee** (IMDCC)	**Stages 5–6, Waste Capture; Monitoring, Research and Development (R&D), Outreach**	The 2006 Marine Debris Act was amended in 2012 and further amended in 2018 and 2020 by the Save Our Seas Act (SOS) and Save Our Seas 2.0 Act (SOS 2.0). Establishes a program at NOAA to "identify, determine sources of, assess, prevent, reduce, and remove marine debris and address the adverse impacts of marine debris on the economy of the United States, marine environment, and navigation safety." Defines "marine debris" to include "any persistent solid material that is manufactured or processed and directly or indirectly, intentionally or unintentionally, disposed of or abandoned into the marine environment or the Great Lakes." (33 U.S.C. § 1956) Amended in 2012 to require NOAA to address marine debris resulting from natural disasters and severe weather events.	Does not establish plastic waste limits or establish liability. Authorizes federal and international coordination through U.S. Coast Guard (USCG), U.S. State Department, U.S. EPA, and IMDCC, the federal interagency coordinating body responsible for addressing marine debris and recommending priorities and strategies, both nationally and internationally. 2020 amendments (SOS 2.0) • Defines and promotes "circular economy." • Established innovation prizes and Marine Debris Foundation to support circular economy goals. • Requires numerous scientific reports and studies focusing on microfibers, microplastics, plastic waste, circular polymers, and derelict fishing gear. • Increased international cooperation and engagement in international treaty discussions. • Authorized additional funding and grant programs for recycling and waste infrastructure.

U.S. Law and Key Agencies	Intervention Stage	Key Provisions	Gaps, Roles, and Related Activities
		Plastic-Related Authorities	

Amended in 2018 "to expand work across the US government, most notably with the US Department of State, to engage foreign governments, especially those of high marine debris source countries, to better address marine debris through strengthened solid waste management. The 2018 Act also mandated that the US government consider addressing marine debris in future trade agreements."[a]

Amended in 2020 to require U.S. EPA to "develop a strategy to improve post-consumer materials management and infrastructure for the purpose of reducing plastic waste and other post-consumer materials in waterways and ocean."

(33 U.S.C. § 4281)

Amended in 2020 requiring IMDCC to develop standardized definitions for the term "microfiber" and U.S. EPA to develop a definition of "microplastics" and standard methodologies to assess and test for the prevalence of microfibers in the ocean and microplastics in drinking water.

Amendments of 2020 also require numerous studies, including this one, as well as a U.S. EPA study with IMDCC and NIST on "minimizing the creation of new plastic waste."

NOAA Program Components[b]:

"(1) identify, determine sources of, assess, prevent, reduce, and remove marine debris;

(2) conduct regional coordination;

(3) reduce adverse impacts of lost and discarded fishing gear, through

- R&D of alternatives and gear marking and recovery techniques; and
- non-regulatory incentives to reduce gear in the environment.

(4) conduct outreach and education;

(5) develop interagency plans in response to 'severe marine debris events,'

- lead coordination with states, tribes, and other federal agencies;
- assess debris composition, volume, and trajectory; and
- estimate potential impacts.

(6) enter into cooperative agreements and contracts and provide financial assistance in the form of grants for projects that address the adverse impacts of marine debris;

(7) reactivate the Interagency Marine Debris Coordinating Committee (IMDCC); and

(8) develop a federal marine debris information clearinghouse."

continued

Microbead Free Waters Act of 2015 **U.S. Food and Drug Administration (U.S. FDA)**	**Stages 1 and 3, Production, Waste Generation**	Amends the Federal Food, Drug, and Cosmetic Act to require the U.S. FDA to prohibit the manufacturing, packaging, and distribution of rinse-off cosmetics containing plastic microbeads. Also applies to products that are both cosmetics and non-prescription (also called "over-the-counter" or "OTC") drugs, such as toothpastes. Defines the term "plastic microbead" as "any solid plastic particle that is—5 millimeters or less in size, and intended to be used to exfoliate or cleanse the body or any part of the body"[c]	Prohibits manufacture of cosmetics containing plastic microbeads but does not prohibit production of plastic microbeads. Congressional and industry support for enactment came in reaction to the rise of multiple state laws banning products containing microbeads.
Solid Waste Disposal Act (SDWA), 42 U.S.C. § 6901 et seq. **Resource Conservation and Recovery Act (RCRA)**	**Stages 3–4, Waste Generation, Management; R&D**	RCRA charges U.S. EPA[d] to set national goals: • "Protecting human health and the environment from the potential hazards of waste disposal. • Conserving energy and natural resources. • Reducing the amount of waste generated. • Ensuring that wastes are managed in an environmentally-sound manner."	Plastic waste, including pellets, currently not defined as a special waste category (e.g., e-waste), a hazardous waste, or meeting RCRA "endangerment" finding, though certain chemicals or additives in plastics may be regulated separately.

U.S. Law and Key Agencies	Intervention Stage	Key Provisions	Gaps, Roles, and Related Activities
		Plastic-Related Authorities	
U.S. EPA **U.S. Department of Commerce (International Trade Administration [ITA], NIST)**		For non-hazardous solid waste, regulations are implemented by states and/or at the local level, with state or local governments given the option to establish more stringent standards. Facilities that do not meet these standards are considered open dumps that must close. "U.S. EPA regulates • Disposal of nonhazardous solid waste implemented by state agencies. • Management of hazardous solid waste. U.S. EPA manages 'hazardous waste from cradle to grave' • Listed hazardous waste; • Characteristic hazardous waste (ignitability, reactivity, corrosivity, toxicity)." Civil liability: Section 7002(a)(1)(B) of RCRA: "any person may commence a civil action against: any person . . . including any past or present generator, past or present transporter, or past or present owner or operator of a treatment, storage, or disposal facility, *who has contributed or who is contributing to the past or present handling, storage, treatment, transportation, or disposal of any solid or hazardous waste which may present an imminent and substantial endangerment to health or the environment* . . . ." 42 U.S.C. § 6972(a)(1)(B)	Nonhazardous solid waste "litter" (plastic waste "leakage" from municipal and other solid waste streams) becomes a municipal enforcement issue. "Disposal" is "the discharge, deposit, injection, dumping, spilling, leaking, or placing of any solid waste . . . into or on any land or water so that such solid waste or hazardous waste or any constituent thereof may enter the environment or be emitted into the air or discharged into any waters, including ground waters." 42 U.S.C. § 6903(3). "Solid waste" as "any garbage, refuse, sludge from a waste treatment plant, water supply treatment plant, or air pollution control facility and other discarded material, including solid, liquid, semisolid, or contained gaseous material resulting from industrial, commercial, mining, and agricultural operations, and from community activities . . . ." 42 U.S.C. § 6903(27) (U.S. EPA 2014).

continued

Authorizes U.S. EPA to implement RCRA's conservation mandate through "sustainable materials management (SMM), a systemic approach to using and reusing materials more productively and effectively over their entire life cycles," primarily implemented at the state and local levels.

- "Requires the Secretary of Commerce to encourage greater commercialization of proven resource recovery technology by stimulating the development of markets for recyclables." Implementation is currently through ITA (global markets) and NIST (standards and research).[g]
- SMM Strategic Plan (2015) [SOS 2.0 requires an updated strategy by end of 2021].[h]
- RCRA SMM Procurement Guidelines (guidelines for federal agencies to procure recyclable items).[i]
- U.S. National Recycling Goal (2020)[j]: Increase national recycling rate to 50% by 2030 by reducing contamination, increasing efficiency, and strengthening recycling markets.
- Draft National Recycling Strategy (2020).[k]
- SOS 2.0 (2020) authorized $55 million/year from 2021 to 2025 for grants to states to implement post-consumer materials management programs (e.g., recycling).

Relevant Civil Case: Charleston Waterkeeper v Frontier Logistics (District Court of South Carolina)[f]: Complaint brought under SDWA/RCRA and the Clean Water Act (CWA) for plastic pellet releases into estuary. Defendant settled in 2021. Complaint: Asserted the company was responsible for "past or present handling, storage, treatment, transportation, or disposal of **solid waste which may present an imminent and substantial endangerment to health or the environment** in violation of RCRA (and discharging without a CWA NPDES permit)." Based "endangerment" claim on lethal and non-lethal effects on wildlife from ingesting plastic pellets (U.S. EPA 2014).

U.S. Law and Key Agencies	Intervention Stage	Key Provisions	Gaps, Roles, and Related Activities
		Plastic-Related Authorities	
Comprehensive Environmental Research, Compensation, and Liability Act (CERCLA), 42 U.S.C. § 9601 et seq. (aka "Superfund") **U.S. EPA** **NOAA (Trustee, Response)**	**Stage 4, Waste Management**	"CERCLA provides U.S. EPA "with the authority to compel responsible parties to **respond to, and remediate, releases of 'hazardous substances'** from facilities and vessels, and addresses **'pollutants or contaminants' posing 'imminent and substantial endangerment.'"** • Responsible parties are defined to include owners and operators of vessels or facilities; transporters; arrangers for disposal. • "Release" is defined to include "leaching," a potential basis for CERCLA action."	Plastic waste is not currently defined as a pollutant, contaminant, or hazardous substance under CERCLA. CERCLA can be used for cleanup of marine debris that contains hazardous substances (e.g., derelict vessels) and to assess threats from releases to human health and the environment.[1]
CWA (33 U.S.C. § 1251, et seq.) and **Rivers and Harbors Appropriation Act of 1899** (33 U.S.C. Chapter 9, § 401 et seq.)	**Stage 4–5 Waste Management, Capture; R&D, Monitoring, Outreach**	CWA provides U.S. EPA with the authority to prohibit point source and indirect pollutant discharges to waters of the United States except under the Act (e.g., with a permit). "Pollutants" are defined broadly and include solid waste, garbage, sewage and sewage sludge, and municipal and agricultural waste (vessel discharges not included, but regulated under section 312).	Prohibits discharging a pollutant from a "point source" except in compliance with the Act (e.g., obtaining a National Pollution Discharge Elimination System [NPDES] permit). Plastics are not defined as a pollutant; if plastic discharges from a facility are addressed at all in NPDES permits, it has been through TSS limits (usually acceptable discharges of visible plastic)

U.S. EPA
NOAA (Trustee, Response)
USCG (Vessel, Response)

Sets water quality, technology, and environmental toxicity standards applicable to industrial and other facilities (e.g., publicly owned treatment works) for

• "Conventional pollutants including biochemical oxygen demand (BOD), total suspended solids (TSS), fecal coliform, pH, and oil and grease;
• Toxic pollutants including 65 pollutants and classes, with 126 specific substances designated as priority pollutants; and
• Nonconventional pollutants."[m]

Requires states to establish water quality standards for every body of water in the state and specify maximum concentrations of pollutants according to water body use.

Requires CWA permit for stormwater or nonpoint source runoff from certain industrial and municipal storm sewer discharges.

Section 303(d) allows for state identification of impaired waters under CWA and Total Maximum Daily Loads (TMDLs) for point and nonpoint sources[n]—can include impairment by "trash." Unless planned measures can be taken to address such impairments, states or U.S. EPA must develop TMDLs for those pollutants.

NPDES permits are also required for nonpoint source (runoff) from certain industrial and municipal systems (often operate under general permits).

CWA programs are delegated to states that meet federal standards; states authorized to set water quality standards for state waters.

TMDLs for trash-impaired waters exist in California, Hawaii, and Alaska.

U.S. EPA's Trash Free Seas program issued new TMDL guidance in 2021: (1) Trash Free Waters (TFW) Stormwater Permit Compendium[o] and (2) U.S. EPA's Escaped Trash Assessment Protocol (ETAP).[p]

U.S. EPA Water Quality Monitoring and Reporting:
• National Water Quality Monitoring Council and Data.[q]
• Section 319 National Nonpoint Source Monitoring Program–U.S. EPA.[t]
• National Coastal Condition Reports–U.S. EPA.[u]

continued

U.S. Law and Key Agencies	Intervention Stage	Key Provisions	Gaps, Roles, and Related Activities
		Plastic-Related Authorities	
		Municipal Separate Storm Sewer System (MS4) permit "requires permittees to develop and implement a comprehensive Storm Water Management Program (SWMP) that must include pollution prevention measures, treatment or removal techniques, monitoring, use of legal authority, and other appropriate measures to control the quality of storm water discharged to the storm drains and thence to waters of the United States." In addition, a small number of municipal governments have set TMDLs limits for trash. Trash Free Waters Program[s] is a voluntary program that provides NPS and other grants to state and local watersheds to address trash and other pollution. The CWA also establishes U.S. EPA and USCG authority for pollution prevention, contingency planning, and response activities within U.S. waters for oil and hazardous substances. Section 312 regulates sewage discharges from vessels; it was amended in 2006 to include the Vessel Incidental Discharge Act. Authorities are implemented by U.S. EPA and USCG (EEZ).	Section 312 Vessel Sewage Discharges: Statutes, Regulations, and Related Laws and Treaties–U.S. EPA.[v] U.S. EPA Research: • U.S. EPA Office of Research and Development: currently conducting research on microplastics. • Region 9: Developing water quality monitoring methods and ASTM standards for sampling microplastics. **Related CWA Decisions and Petitions:** *Formosa Permit and Decision*[w] "Formosa's 2016 Permit prohibits the 'discharge of floating solids or visible foam in other than trace amounts' ... Moreover, TCEQ rules prohibit the discharge of 'floating debris and suspended solids' into surface waters. [. . .] The undisputed evidence shows that plastic pellets and PVC powder discharged by Formosa caused or contributed to the damages suffered by the recreational, aesthetic, and economic value of [surface water]. . . ."

		Authorizes a range of waste water infrastructure grants (wastewater treatment and nonpoint source) • Section 10 of the Rivers and Harbors Appropriation Act requires a permit to be issued by the U.S. Army Corps of Engineers for discharges of any dredged or fill material (including plastics) into the navigable waters of the United States.	*June 2019 Petition Under 40 CFR Parts 414 and 419*[x]: requested, inter alia, "prohibit the discharge of plastic pellets and other plastic materials in industrial stormwater and wastewater" One case alleged plastic pellets are a "pollutant" under the CWA. Settled; not adjudicated.[f]

Clean Air Act (CAA), 42 U.S.C. § 7401 et seq.
 	
U.S. EPA
	Stage 4, Waste Management; Monitoring	"CAA regulates ambient air quality by limiting sources of air pollutants from
		• Stationary sources of criteria pollutants to meet air quality and technology standards, as well as hazardous air pollutants:
		  ○ Criteria pollutants include nitrogen and sulfur dioxides, carbon monoxide, lead, and ozone as volatile organic compounds, and particulate less than 2.5 micrometers;
		  ○ Hazardous air pollutants—187 chemicals listed for carcinogenicity, toxicity, and other potential harms.
		• Mobile sources of criteria air pollutants from internal combustion engines."[e]
			"Microplastics air emissions from ground level sources are generally not covered—leaving no path to directly limit these microplastic emissions to ambient air. Plastic component of PM 2.5 appears to be difficult to completely capture and analyze due to limitations in sampling and analytical methods."[e]
			
			In 2021, U.S. EPA announced it is considering more stringent regulation of pyrolysis and gasification (sometimes used in plastic chemical recycling) under CAA § 129.[y]

continued

U.S. Law and Key Agencies	Intervention Stage	Key Provisions	Gaps, Roles, and Related Activities
		Plastic-Related Authorities	
Safe Drinking Water Act (SDWA), 42 U.S.C. § 300f et seq **U.S. EPA**	**Stage 4 Waste Management; Stage 5, Waste Capture; Monitoring**	"SDWA regulates public water supply, imposing maximum contaminant limits for chemical contaminants including • Microorganisms and viruses, turbidity (cloudiness, suspended solids) up to 0.3 ntu [nephelometric turbidity unit], disinfectants, disinfection byproducts, inorganic chemicals, organic chemicals (not plastics), radionuclides. • Also, monitoring for unregulated contaminants (including perflurooctanoic acid [PFOA] and perflurooctane sulfonate [PFOS]). Consumer confidence reports and public notifications."[e] U.S. EPA provides grants to implement state drinking water programs and to help each state set up a special fund to assist public water systems in financing the costs of improvements (called the Drinking Water State Revolving Fund).[z]	"Microplastics are not included unless captured as Turbidity—allowed up to 0.3 Nephelometric Turbidity Units."[e] SOS 2.0 (33 U.S.C. § 4282—Grant programs) clarified that SDWA infrastructure grants can be used to "support improvements in reducing and removing plastic waste, including microplastics and microfibers, from drinking water."
Toxic Substances Control Act (TSCA), 15 U.S.C. § 2601 et seq. (1976) **U.S. EPA**	**Stage 3 Waste Generation; Stage 4, Waste Management; R&D**	Provides U.S. EPA with "authority to require reporting, record-keeping and testing requirements, and restrictions relating to chemical substances and/or mixtures" (does not include food, drugs, cosmetics, and pesticides). "TSCA can potentially be used for the purpose of addressing risks specific to chemical substances that may be in plastic waste."[aa]	U.S. EPA has not used these authorities to regulate plastic waste. "40 CFR 723.250(d) Polymer 40 CFR 723.250(d)—exempts from Premanufacture Notice requirements those polymers that are inert: (1) based on level of concern regarding functional groups, and (2) not excluded from the exemption.

Ocean Dumping Act (Marine Protection Research and Sanctuaries Act of 1972), as amended by the Ocean Dumping Ban Act of 1988 and the Water Resources Development Act of 1992 **NOAA** **U.S. EPA** **USCG** **U.S. Army Corps of Engineers**	**Stage 6, Minimize at-Sea Disposal; Monitoring**	1972 law, as amended, implemented by the USCG, U.S. EPA, and the U.S. Army Corps of Engineers— • "Prohibits the ocean dumping of municipal sewage sludge and industrial wastes, such as wastes from plastics and pharmaceutical manufacturing plants and from petrochemical refineries."[dd] • Bans the ocean disposal of "medical waste"[dd] (1988). • "Ocean dumping permits, including for ocean disposal of dredged material, conform to long-term management plans to ensure that permitted activities are consistent with expected uses of designated ocean disposal sites"[dd] (1992).	Prohibitions align with the requirements of international law under the 1972 London Dumping Convention (Convention on the Prevention of Marine Pollution by Dumping of Wastes and Other Matter of 1972) and its successor, the 1996 London Protocol (in effect starting 2006). The United States has not ratified the Convention or Protocol but does participate in meetings (Secretariat at the International Maritime Organization). Any materials dumped in the ocean are "evaluated to ensure that they will not pose a danger to human health or the environment and that there are no better alternatives for their reuse or disposal."[dd]

Still covering polymers that are cationic, degradable or unstable, water-absorbing, or vulnerable to reactants. The more inert the substance is, the less regulated it is."[e] "Lautenberg Chemical Safety Act of 2016 amended TSCA to require EPA to evaluate existing chemicals with clear and enforceable deadlines; conduct risk-based chemical assessments; increase public transparency for chemical information."[bb] In 2021, several chemicals used in plastic and rubber manufacturing were added to list of chemicals regulated under TSCA.[cc]

continued

U.S. Law and Key Agencies	Intervention Stage	Key Provisions	Gaps, Roles, and Related Activities
		Plastic-Related Authorities	
Act to Prevent Pollution from Ships as amended by Marine Plastic Pollution Research and Control Act (MARPOL) 33 U.S.C. § 1901 et seq. **USCG** **NOAA**	**Stage 6, Minimize at-Sea Disposal; Monitoring**	• "Implements the provisions of Annex V of the International Convention for the Prevention of Pollution from Ships (MARPOL) relating to garbage and plastics."*ee* "Applies to all vessels, whether seagoing or not, regardless of flag, on the navigable waters of the U.S. and in the exclusive economic zone of the U.S. It applies to U.S. flag vessels wherever they are located."*ff* • Prohibits the "discharge of plastics, including synthetic ropes, fishing nets, plastic bags, and biodegradable plastics, into the water."*ff* • Prohibits "discharge of floating dunnage, lining, and packing materials in the navigable waters and in areas offshore less than 25 nautical miles from the nearest land."*ff* • "Ships" includes fixed or floating platforms, which are subject to separate garbage discharge provisions. "For these platforms, and for any ship within 500 meters of these platforms, disposal of all types of garbage is prohibited."*ff*	Other • "Food waste or paper, rags, glass, metal, bottles, crockery and similar refuse cannot be discharged in the navigable waters or in waters offshore inside 12 nautical miles from the nearest land."*gg* • "Food waste, paper, rags, glass, and similar refuse cannot be [discharged] in the navigable waters or in waters offshore inside three nautical miles from the nearest land (some exceptions for emergencies)."*gg* • Requires "all manned, oceangoing U.S. flag vessels of 12.2 meters or more in length engaged in commerce, and all manned fixed or floating platforms subject to the jurisdiction of the U.S., to keep records of garbage discharges and disposals."*gg* See, e.g., Hagen (1990).

| Coastal Zone Management Act, 16 U.S.C. § 1451 et seq. NOAA | Stage 4, Waste Management | • Establishes National Coastal Zone Management Program, a unique federal-state partnership for management of the coastal zone (including Great Lakes) aimed at protecting, preserving, and enhancing resources of the coastal zone.
• The Coastal Nonpoint Pollution Control Program (Section 6217 of the 1990 amendments) "requires states and territories with approved Coastal Zone Management Programs to develop Coastal Nonpoint Pollution Control Programs"[hh] that lay out management measures. Administered jointly with U.S. EPA. | States receive federal grants to support creation and implementation of coastal zone management plans. Plans approved by the Secretary of Commerce become the state's governing rules for development of the coastal zone. Law provides states the assurance that federal activities (including federally permitted activities) in the coastal zone must be "consistent" with the state plan. Also incentivizes protection of natural resources such as wetlands, control of marine debris, addressing coastal hazards, ocean planning, and energy siting[ii] |
| Federal Trade Commission Act (15 U.S.C. §§ 41-58, as amended)

Federal Trade Commission (FTC) | Stage 2, Innovate Material and Product Design | Section 5 of the Federal Trade Commission Act (15 U.S.C. §§ 45) "prohibits unfair or deceptive practices in or affecting commerce."
• FTC has Green Guides to help companies appropriately address environmental marketing
• Can bring enforcement actions for claims that deviate from the Guides. "A representation, omission, or practice is deceptive if it is likely to mislead consumers acting reasonably under the circumstances and is material to consumers' decisions."[jj] | FTC's Green Guides cover claims such as "recycling" "biodegradable." Last update was 2012, next expected 2022; updates not required by law (see GAO 2020). See 16 C.F.R. pt. 260.

Examples of plastic labeling actions:
Biodegradable
Oxodegradable
Post-consumer recycled plastic content "FTC considers three factors when determining whether a practice is unfair: (1) whether it injures consumers, (2) whether it violates established public policy, and (3) whether it is unethical or unscrupulous." |

continued

U.S. Law and Key Agencies	Intervention Stage	Key Provisions	Gaps, Roles, and Related Activities
		Research, Development, and Monitoring Authorities	
National Energy Policy and Programs 42 U.S.C. 149 (§ 15801 et seq.) **U.S. Department of Energy** (DOE)	**R&D**	Authorities emphasize energy efficiency and innovative materials research.	Plastics Innovation Challenge Draft Roadmap and Request for Information.[kk] The REMADE Institute; 2020 Impact Report.[ll] DOE Bio-optimized Technologies to keep Thermoplastics out of Landfills and the Environment (BOTTLE) Consortium.[mm] DOE American Chemistry Council Memorandum of Understanding 2020 Innovative Plastics Recycling Technologies.[nn]
Integrated Coastal and Ocean Observation System Act of 2009, reauthorized in the Coordinated Ocean Observations and Research Act of 2020; related research authorities. 33 U.S.C. §§ 3601–3610 (Subtitle C) **NOAA,** with 11 other federal agencies	**R&D**	Authorizes a "national integrated System of ocean, coastal, and Great Lakes observing systems (federal and non-federal)" and includes "in situ, remote, and other coastal and ocean observation, technologies, and data management and communication systems, designed to address regional and national needs for ocean information"; authorizes basic and applied research. NOAA also operates weather and climate observations systems (satellite, ground, and air) through related systems under the National Weather Service Organic Act.	Identifies NOAA as lead federal agency for the system and establishes two governance mechanisms: (1) federal coordination by a White House National Ocean Research and Leadership Council, (2) Office of Science and Technology Policy Interagency Ocean Observation Committee (consisting of 12 federal agencies, led by 4 co-chairs). Agencies: NOAA, National Science Foundation (NSF), National Aeronautics and Space Administration (NASA), U.S. EPA, Bureau of Ocean Energy Management (BOEM), Marine Mammal Commission (MMC), Joint Chiefs, Office of Naval Research, U.S. Army Corps of Engineers, USCG, U.S. Geological

		Survey (USGS), U.S. Department of State. Affiliated with 11 U.S. Regional Observing Systems, and coordinates with the Global Ocean Observing System. Connects with NOAA's weather and climate observation systems.
U.S. Geological Survey Organic Act 43 U.S.C. Chapter 2 (§ 31 et seq.) **USGS**	R&D	Authorizes USGS to examine the geological structure, mineral resources, and products of the national domain, which provides scientific information to, among other things, understand Earth systems, manage water, and enhance and protect quality of life. Operates the USGS Stream Gauge Network, consisting of more than 11,000 gauges that collect data for a variety of uses, including by other agencies, including hazard and flood information as well as assessing water quality, regulating point source discharges, assessing if streams are safe for recreational activities.
Other Ocean R&D **NSF** (42 USC Chapter 16) **NASA Organic Act** 42 U.S.C. § 2451 et seq. **Oceans and Human Health Act (NOAA),** 33 U.S.C. Chapter 44, Sec. 3101 et seq.	R&D	Provides authority for NSF to support scientific research and education. NASA is authorized to conduct scientific research, measurement, monitoring, and outreach related to aeronautical and space activities (e.g., remote sensing). NOAA is authorized to conduct and support a range of atmospheric, ocean, and coastal research authorities, including research relevant to ocean and human health NSF has provided funding for a number of ocean plastic and materials research projects under a range of grant programs. NASA operates satellite remote sensing and related research efforts relevant to ocean conditions and constituents. NOAA research relates to ocean health and ecosystem conditions, beyond the purview of the Marine Debris Act.

continued

U.S. Law and Key Agencies	Intervention Stage	Key Provisions	Gaps, Roles, and Related Activities
		Research, Development, and Monitoring Authorities	

Non-statutory: U.S. (federal and state)
Common Law
 	
Legal claims require proof of "injury" to meet "standing" requirements
		Lujan v. Defenders of Wildlife - 504 U.S. 555, 112 S. Ct. 2130 (1992)[aa]
		• "Plaintiffs must have "standing"
		  ○ "injury in fact, which means an invasion of a legally protected interest that is concrete and particularized, and actual or imminent, not conjectural or hypothetical;"
		  ○ "a causal relationship between the injury and the challenged conduct, i.e., the injury can be fairly traced to the challenged action of the defendant, and has not resulted from the independent action of some third party not before the court;" and
		  ○ "a likelihood that the injury will be redressed by a favorable decision, which means that the prospect of obtaining relief from the injury as a result of a favorable ruling is not too speculative."
		• Claims must be ripe and not moot.
		• Necessary and indispensable parties.
		• Burden of proof in civil cases is "preponderance of the evidence."
			For example, "Earth Island v. Crystal Geyser, Clorox, Coca-Cola, Pepsico, Nestle, Mars, et al. (2020, Filed Sup Ct CA): Misleading claims of recyclability. Federal court remanded to state court 2021claims under
			• CA Consumers Legal Remedies Act
			• Public Nuisance
			• Breach of Express Warranty
			• Strict Liability-Failure to Warn
			• Negligence and Negligence–Failure to Warn"[e]

[a]See https://www.env.go.jp/press/files/en/872.pdf.

[b]See https://marinedebris.noaa.gov/sites/default/files/mdp_pea.pdf.

[c]See https://www.fda.gov/cosmetics/cosmetics-laws-regulations/microbead-free-waters-act-faqs.

[d]See https://www.epa.gov/history/epa-history-resource-conservation-and-recovery-act.

[f]See Ternes and Fulton (2020).

[f]See https://www.southernenvironment.org/wp-content/uploads/legacy/words_docs/FINAL_COMPLAINT.pdf.

[g]See https://www.gao.gov/assets/gao-21-87.pdf.

[h]See https://www.epa.gov/sites/production/files/2016-03/documents/smm_strategic_plan_october_2015.pdf.

[i]See https://www.epa.gov/smm/comprehensive-procurement-guideline-cpg-program#bio.

[j]See https://www.epa.gov/americarecycles/us-national-recycling-goal.

[k]See https://www.epa.gov/americarecycles/draft-national-recycling-strategy-and-executive-summary.

[l]See Kimrey and Helton (2014).

[m]See https://www.epa.gov/eg/learn-about-effluent-guidelines.

[n]See https://www.epa.gov/tmdl/overview-total-maximum-daily-loads-tmdls#2.

[o]See https://www.epa.gov/trash-free-waters/trash-stormwater-permit-compendium.

[p]See https://www.epa.gov/trash-free-waters/epas-escaped-trash-assessment-protocol-etap.

[q]See https://www.waterqualitydata.us/.

[r]See https://www.epa.gov/tx/municipal-separate-storm-sewer-system-ms4-storm-water-management-program-swmp.

[s]See https://www.epa.gov/trash-free-waters.

[t]See https://www.epa.gov/nps/national-nonpoint-source-monitoring-program.

[u]See https://www.epa.gov/national-aquatic-resource-surveys/national-coastal-condition-reports.

[v]See https://www.epa.gov/vessels-marinas-and-ports/vessel-sewage-discharges-statutes-regulations-and-related-laws-and.

[w]See *San Antonio Bay Estuarine Waterkeeper v. Formosa Plastics Corp.*, civil action no. 6:17-CV-0047 (S.D. Tex. Jun. 27, 2019).

[x]See https://waterkeeper.org/wp-content/uploads/2019/07/CWA-Petro-Plastics-Petition-to-EPA-6-23-19.pdf.

[y]See https://www.govinfo.gov/content/pkg/FR-2021-09-03/pdf/2021-19390.pdf.

[z]See https://www.epa.gov/dwsrf.

[aa]See https://g20mpl.org/partners/unitedstates.

[bb]See https://www.epa.gov/assessing-and-managing-chemicals-under-tsca/frank-r-lautenberg-chemical-safety-21st-century-act.

[cc]See https://www.epa.gov/assessing-and-managing-chemicals-under-tsca/persistent-bioaccumulative-and-toxic-pbt-chemicals.

[dd]See https://www.epa.gov/ocean-dumping/learn-about-ocean-dumping.

[ee]See https://www.epa.gov/environmental-topics/water-topics#our-waters.

[ff]See NOAA (1998).

[gg]See https://coast.noaa.gov/data/docs/nerrs/Reserves_ACE_SiteProfile.pdf.

[hh]See Cody et al. (2012).

[ii]See https://coast.noaa.gov/czm/.

[jj]See U.S. GAO. 2020. Recycling: Building on Existing Federal Efforts Could Help Address Cross-Cutting Challenges. United States Government Accountability Office. https://www.gao.gov/products/gao-21-87.

[kk]See https://www.energy.gov/plastics-innovation-challenge/downloads/plastics-innovation-challenge-draft-roadmap-and-request.

[ll]See https://remadeinstitute.org/.

[mm]See https://www.bottle.org/index.html.

[nn]See https://www.energy.gov/articles/us-energy-department-and-american-chemistry-council-sign-memorandum-understanding.

[oo]See https://www.lexisnexis.com/community/casebrief/p/casebrief-lujan-v-defenders-of-wildlife, as presented in Ternes and Fulton (2020).

"there were 331 local plastic bag ban ordinances across 24 states in the United States."[2] However, some municipalities have been challenged by state preemption laws and lawsuits around local ordinances.

Laws generally fall into the following categories:

- Single-use bans and fees,
- Extended producer responsibility,
- Bottle bills,
- Per-and polyfluoroalkyl substances (PFAS) in packaging, and

Task force and study commissions.[3,4]

REFERENCES

Cody, B., J. Schneider, M. Tiemann, and G. Relf. 2012. *Selected Federal Water Activities: Agencies, Authorities, and Congressional Committees.* Congressional Research Service. http://nationalaglawcenter.org/wp-content/uploads/assets/crs/R42653.pdf.

G20. 2021. "The United States: Actions and Progress on Marine Plastic Litter." https://g20mpl.org/partners/unitedstates.

Hagen, P. 1990. "The international community confronts plastics pollution from ships: MARPOL Annex V and the problem that won't go away." *Am Univ Int Law Rev* 5 (2):425-496.

Kimrey, C. and D. Helton. 2014. "Abandoned vessel authorities and best practices guidance – A review of NRT work". *International Oil Spill Proceedings.* 1:2053–2063. doi: 10.7901/2169-3358-2014.1.2053.

Ministry of Environment, Japan. 2020. *G20 Report on Actions Against Marine Plastic Litter: Second Information Sharing Based on the G20 Implementation Framework.* Tokyo, Ministry of Environment, Japan. https://www.env.go.jp/press/files/en/872.pdf.

NOAA (National Oceanic and Atmospheric Administration). Office of the Chief Scientist. 1998. Year of the Ocean discussion papers. Washington, D.C.: NOAA.

Ternes, M. E., and S. Fulton. 2020. Overview of United States Law Governing Solid and Water Waste Management. Presentation at an online meeting of the Committee on the U.S. Contributions to Global Ocean Plastic Waste, December 7.

U.S. EPA (U.S. Environmental Protection Agency). 2014. *RCRA's Critical Mission & the Path Forward.* Washington D.C: U.S. EPA. https://www.epa.gov/rcra/resource-conservation-and-recovery-act-critical-mission-path-forward.

U.S. GAO. 2020. Recycling: Building on Existing Federal Efforts Could Help Address Cross-Cutting Challenges. United States Government Accountability Office.

---

[2]See https://law.ucla.edu/news/federal-actions-address-marine-plastic-pollution.

[3]See https://www.ncelenviro.org/issue/plastic-pollution/.

[4]Other summaries of federal and state activities related to marine debris and plastic waste: https://law.ucla.edu/news/federal-actions-address-marine-plastic-pollution; https://www.epa.gov/environmental-topics/land-waste-and-cleanup-topics; and https://www.ncsl.org/research/environment-and-natural-resources/plastic-bag-legislation.aspx.

D

Estuary Table

TABLE D-1 Peer-reviewed Studies in Which Plastic Waste Was Measured in Estuaries and Rivers of the United States

Study	Locale	Sampling Dates	Environmental Matrix (N = number of sites)
Moore, Lattin, and Zellers 2011	Los Angeles and San Gabriel Rivers	Two occupations: Nov 22 or Dec 28, 2004, and Apr 11, 2005	Surface, mid-depth, and bottom samples (N = 3, two occupations)
Yonkos et al. 2014	Chesapeake Bay	~Monthly between July and Dec 2011	4 estuarine tributaries, surface water (N = 60)
Bikker et al. 2020	Chesapeake Bay	Single occupation collected Aug 31–Sep 18, 2015	Estuary surface water (N = 30)
Davis and Murphy 2015	Salish Sea & Inside Passage (WA)	2011 (N = 62), 2012 (N = 15)	Estuary surface water (N = 77)
McCormick et al. 2016	9 rivers in Chicago metropolitan area (IL, IN)	Single occupation collected July 10–Oct 13, 2014	Stream surface water (N = 9, each site with 4 replicates at both locations upstream and downstream of wastewater treatment plant [WWTP] outfall site)
Hoellein et al. 2017	North Shore Channel (urban waterway, Chicago, IL)	Aug 7, 2017	Channel surface water and benthic sediment (N = 5, 4 replicates of each sample type at each location)
Baldwin, Corsi, and Mason 2016	29 Great Lakes tributaries (6 states)	Apr 2014–Apr 2015, each tributary sampled 3–4 times	Surface water (N = 107)
Sutton et al. 2016	San Francisco Bay	Single occupation collected on 2 days in Jan 2015	Estuary surface water (N = 9)
Sutton et al. 2019	San Francisco Bay and Tomales Bay	Two occupations (wet/dry conditions)	Estuary surface water (N = 17) and sediments (N = 20)
Miller et al. 2017	Hudson River (NY)	Single occupation collected in June and Oct 2016	River surface water (N = 142)

Sampling Method	Abundance, as Reported	Notes
Manta net and hand nets (0.333- to 0.8-mm mesh)	0 to 12,932 particles/m^3; 0 to 121 g/m^3	Sampled during dry period (Nov/Dec) and within 24 hours of 0.25 in. of rainfall (Apr)
Manta net (0.3-mm mesh)	From <1.0 to >560 g/km^2	Peaks in abundance after major storm events
Manta net (0.33-mm mesh)	0.007 to 1.245 particles/m^3	Not all particles were plastic Polyethylene (PE), polypropylene (PP) most common plastics found
Manta net (0.335-mm mesh)	0 to >130,000 particles/km^2	Samples dominated by expanded polystyrene (EPS) foam
Neuston net (0.333-mm mesh)	0.48 (±0.09) to 11.22 (±1.53) particles/m^3	Highly variable particle flux between sites; mainly PE, PP, polystyrene (PS); 7 of 9 sites had higher concentrations downstream of WWTP effluent
Neuston net (0.333 mm), Ponar grab (~0.75–1 L sediment)	1.67 particles/m^3 to 10.36 particles/m^3 (water); 36 to 1,613 particles/L (sediment)	Much higher microplastic abundance in sediment than in surface water; microplastic abundance in water did not vary with increasing distance downstream of WWTP outfall
Neuston net (0.333-mm mesh)	0.05 to 32 particles/m^3	Plastics found in all samples. Majority were fibers/lines whose concentrations were not related to watershed attributes or hydrological processes
Manta net (0.333-mm mesh)	15,000 to 2,000,000 particles/km^2	Abundances higher in southern bay than central bay
Manta net (0.335-mm mesh), 1-L water grab sample, pumped water sample, sediment grab	2,400 to 6,200,000 particles/km^2 in surface water; 0.5 to 60 particles/g dry weight	Abundances include microplastics and other microparticles. Surface water samples collected in the wet season had higher concentrations of microplastics than in the dry season.
Water grab samples, filtered on 0.45-µm filter	0 to 12.37 microfibers/L	Abundances include microplastics and other microfibers

continued

TABLE D-1 Continued

Study	Locale	Sampling dates	Environmental matrix (N = # of sites)
Gray et al. 2018	Charleston Harbor, Winyah Bay (SC)	Single occupation	Sea surface microlayer (N = 12), intertidal sediment (N = 10)
Barrows et al. 2018	Gallatin River basin (MT, WY)	Sept 2015–June 2017	River surface water (N = 714, occupied seasonally over 2 years at 72 sites)
Kapp and Yeatman 2018	Snake River (WY, ID, OR, WA)	5 repeated sampling periods between June and Aug 2015	River surface water
Cohen et al. 2019	Delaware Bay	Apr 21, 28, 2017, and June 12, 13, 2017	Estuary surface water (N = 16, occupied once in Apr and once in June)
McEachern et al. 2019	Tampa Bay (FL)	1–5 months between samples from June 2016 to July 2017 (water); Single occupation, Mar 21–23, 2017 (sediment)	Surface water (N = 24; 2 methods), sediment (N = 9)
Lenaker et al. 2019	Milwaukee River Basin	5 sampling trips, May to Sept 2016 (water sampling); June 2016 (sediment)	Stream/river/estuary surface water and subsurface water (N = 96), sediment (N = 9)
Christensen et al. 2020	Blacksburg, VA region	Single occupation on June 21, 2018 and Aug 31, 2018	River bed, banks, and floodplain sediment from 3 rivers (N = 14)
Bailey et al. 2021	Raritan River and Raritan Bay (NJ)	July 26, 2018 (low flow), Apr 11, 2019 (moderate flow), Apr 16, 2019 (high flow)	River and estuary surface water (N = 14, some duplicates)

Sampling method	Abundance, as reported	Notes
4-L sea surface microlayer samples; top 2 cm of sediment in quadrats	0 to 1195.7 ± 193.9 particles/m^2 in sediment; 3 to 88 particles/L in water	High abundance of suspected tire wear particles (Charleston Harbor)
~1-L water samples	0 to 67.5 particles/L	Majority of particles were fibers (80%); microplastic concentration inversely related to river discharge
1.85-L water samples (N = 28); net samples (0.100-mm mesh) (N = 28)	0 to 5.405 particles/L (bulk water samples); 0 to 13.701 particles/m^3 (net samples)	
Ring plankton net (0.2-mm mesh)	0.19 to 1.24 particles/m^3	High spatial/temporal variability
1-L water samples; plankton net (0.33-mm mesh); Shipek grab for sediment	0 to 7.0 particles/L (bulk water samples); 1.2 to 18.1 particles/m^3 (net tow samples); 30 to 790 particles/kg (sediment)	
Neuston net (0.333-mm mesh); Circular net (0.333-mm) for subsurface; spoons for sediment	0.21 to 19.1 particles/m^3 at surface; 0.06 to 4.3 particles/m^3 subsurface; 32.9 to 6,229 particles/kg dry weight sediment	Concentration of low-density particles decreased with depth; concentration of high-density particles increased with depth
Hand trowel (40 cm × 40 cm area × 4 cm depth)	Averages by site ranged from 17 particles/kg to 180 particles/kg;	Average concentration was as high or higher in floodplain than in stream channel, and average particle size was also larger
Plankton net (0.080- or 0.150-mm mesh)	0 to 2.75 particles/m^3 for 500–2,000 μm size class; 0.38 to 4.71 particles/m^3 for 250–500 μm size class	

REFERENCES

Bailey, K., K. Sipps, G. K. Saba, G. Arbuckle-Keil, R. J. Chant, and N. L. Fahrenfeld. 2021. "Quantification and composition of microplastics in the Raritan Hudson Estuary: Comparison to pathways of entry and implications for fate." *Chemosphere* 272. doi: 10.1016/j.chemosphere.2021.129886.

Baldwin, A. K., S. R. Corsi, and S. A. Mason. 2016. "Plastic debris in 29 Great Lakes tributaries: Relations to watershed attributes and hydrology." *Environ Sci Technol* 50 (19):10377-10385. doi: 10.1021/acs.est.6b02917.

Barrows, A. P. W., K. S. Christiansen, E. T. Bode, and T. J. Hoellein. 2018. "A watershed-scale, citizen science approach to quantifying microplastic concentration in a mixed land-use river." *Water Res* 147:382-392. doi: 10.1016/j.watres.2018.10.013.

Bikker, J., J. Lawson, S. Wilson, and C. M. Rochman. 2020. "Microplastics and other anthropogenic particles in the surface waters of the Chesapeake Bay." *Mar Pollut Bull* 156:111257. doi: 10.1016/j.marpolbul.2020.111257.

Christensen, N. D., C. E. Wisinger, L. A. Maynard, N. Chauhan, J. T. Schubert, J. A. Czuba, and J. R. Barone. 2020. "Transport and characterization of microplastics in inland waterways." *J Water Process Eng* 38. doi: 10.1016/j.jwpe.2020.101640.

Cohen, J. H., A. M. Internicola, R. A. Mason, and T. Kukulka. 2019. "Observations and simulations of microplastic debris in a tide, wind, and freshwater-driven estuarine environment: The Delaware Bay." *Environ Sci Technol* 53 (24):14204-14211. doi: 10.1021/acs.est.9b04814.

Davis, W., III, and A. G. Murphy. 2015. "Plastic in surface waters of the Inside Passage and beaches of the Salish Sea in Washington State." *Mar Pollut Bull* 97 (1-2):169-177. doi: 10.1016/j.marpolbul.2015.06.019.

Gray, A. D., H. Wertz, R. R. Leads, and J. E. Weinstein. 2018. "Microplastic in two South Carolina estuaries: Occurrence, distribution, and composition." *Mar Pollut Bull* 128:223-233. doi: 10.1016/j.marpolbul.2018.01.030.

Hoellein, T. J., A. R. McCormick, J. Hittie, M. G. London, J. W. Scott, and J. J. Kelly. 2017. "Longitudinal patterns of microplastic concentration and bacterial assemblages in surface and benthic habitats of an urban river." *Freshwater Sci* 36 (3):491-507. doi: 10.1086/693012.

Kapp, K. J., and E. Yeatman. 2018. "Microplastic hotspots in the Snake and Lower Columbia rivers: A journey from the Greater Yellowstone Ecosystem to the Pacific Ocean." *Environ Pollut* 241:1082-1090. doi: 10.1016/j.envpol.2018.06.033.

Lenaker, P. L., A. K. Baldwin, S. R. Corsi, S. A. Mason, P. C. Reneau, and J. W. Scott. 2019. "Vertical distribution of microplastics in the water column and surficial sediment from the Milwaukee River Basin to Lake Michigan." *Environ Sci Technol* 53 (21):12227-12237. doi: 10.1021/acs.est.9b03850.

McCormick, A. R., T. J. Hoellein, M. G. London, J. Hittie, J. W. Scott, and J. J. Kelly. 2016. "Microplastic in surface waters of urban rivers: Concentration, sources, and associated bacterial assemblages." *Ecosphere* 7 (11). doi: 10.1002/ecs2.1556.

McEachern, K., H. Alegria, A. L. Kalagher, C. Hansen, S. Morrison, and D. Hastings. 2019. "Microplastics in Tampa Bay, Florida: Abundance and variability in estuarine waters and sediments." *Mar Pollut Bull* 148:97-106. doi: 10.1016/j.marpolbul.2019.07.068.

Miller, R. Z., A. J. R. Watts, B. O. Winslow, T. S. Galloway, and A. P. W. Barrows. 2017. "Mountains to the sea: River study of plastic and non-plastic microfiber pollution in the northeast USA." *Mar Pollut Bull* 124 (1):245-251. doi: 10.1016/j.marpolbul.2017.07.028.

Moore, C. J., G. L. Lattin, and A. F. Zellers. 2011. "Quantity and type of plastic debris flowing from two urban rivers to coastal waters and beaches of Southern California." *J Integr Coast Zone Manag* 11 (1):65-73.

Sutton, R., S. A. Mason, S. K. Stanek, E. Willis-Norton, I. F. Wren, and C. Box. 2016. "Microplastic contamination in the San Francisco Bay, California, USA." *Mar Pollut Bull* 109 (1):230-235. doi: 10.1016/j.marpolbul.2016.05.077.

Sutton, R., A. Franz, A. Gilbreath, D. Lin, L. Miller, M. Sedlak, A. Wong, C. Box, R. Holleman, K. Munno, X. Zhu, C. Rochman. 2019. *Understanding Microplastic Levels, Pathways, and Transport in the San Francisco Bay Region.* SFEI Contribution No. 950. Richmond, CA: San Francisco Estuary Institute.

Yonkos, L. T., E. A. Friedel, A. C. Perez-Reyes, S. Ghosal, and C. D. Arthur. 2014. "Microplastics in four estuarine rivers in the Chesapeake Bay, U.S.A." *Environ Sci Technol* 48 (24):14195-202. doi: 10.1021/es5036317.

E

Global Instruments and Activities Relevant to Ocean Plastic Pollution

By 2000, there have been five binding international ocean plastic pollution policies that addressed maritime sources of pollution (Karasik et al. 2020). Since 2000, there have been 28 nonbinding international policies ("soft law") addressing land-based sources (Karasik et al. 2020). "However, there are no agreed-upon global, binding, specific, and measurable targets to reduce plastic pollution" (Karasik et al. 2020). In 2021, there was growing momentum and support for strengthening existing instruments and for the negotiation of a global convention on plastics and plastic pollution.

Global or Regional Organization	Legal Instrument and Relevant Coverage
Focus: Plastic Pollution	
United Nations General Assembly (UNGA)	2012: UNGA Resolution[a] *2015: UNGA Resolution 70/1 Sustainable Development.* Agreed on 2030 Agenda for Sustainable Development and set Sustainable Development Goals (SDGs), which include a target (SDG 14.1) that member states should *"by 2025, prevent and significantly reduce marine pollution of all kinds, in particular from land-based activities, including marine debris and nutrient pollution."* *2021–2022 (Under Discussion): Global Convention on Plastics and Plastic Pollution.* United Nations Environmental Assembly (UNEA) conducting discussions on a "Convention on Plastics and Plastic Pollution" based on the Montreal Protocol (see below); supported also by 2021 G7 communique.
United Nations Environment Programme (UNEP)	*2011: The Honolulu Strategy: A Global Framework for Prevention and Management of Marine Debris[b]* (UNEP, National Oceanic and Atmospheric Administration). Three major goals were: • Goal A: Reduced amount and impact of land-based sources of marine debris introduced into the sea • Goal B: Reduced amount and impact of sea-based sources of marine debris • Goal C: Reduced amount and impact of accumulated marine debris on shorelines, in benthic habitats, and in pelagic waters Three "extremely important" issues were deemed beyond the scope of this strategy and needed to be addressed holistically: • Zero target for marine debris creation • Integrated solid waste management • Extended producer responsibility

United Nations Environmental Assembly (UNEA)

2014: UNEA/Resolution 1/6 "Marine Plastic Debris and Microplastics"

2016: UNEA Resolution 2/11 "Marine Plastic Litter and Microplastics"

2017: Report—Finds the "absence of an institution with a mandate to coordinate existing efforts, lack of legally binding instruments in key regions to manage marine plastic pollution originating from land, and limited industry due diligence and lack of global design standards to mitigate plastic pollution hamper effective international management of plastics."[c]

2018: UNEA Resolution 3/7 "Marine Litter and Microplastics"

2019: UNEA Resolution 4/6 "Member states called for more rigorous monitoring of the status of the global plastic pollution problem and efforts to address it, including existing activities and actions by governments."

2019: UNEA Resolution 4/9 "Addressing Single-Use Plastic Products Pollution"

2021: UNEP Report *Global Assessment of Marine Litter and Plastic Pollution*, published in October 2021 to inform UNEA-5.2.

2021-2022: A special group (Ad Hoc Open-Ended Expert Group) is exploring how to tackle marine plastic pollution. At the 5th session (UNEA-5) member states will discuss "need for negotiations for a new Convention to begin or not, and whether the Ad Hoc Open-Ended Expert Group needs more time to consider governance options."[d] UNEA-5 meetings were held in February 2021 and will be held in February 2022.

continued

Global or Regional Organization	Legal Instrument and Relevant Coverage
Focus: Pollution-Oriented Agreements Relevant to Plastic Pollution and Marine Debris	
United Nations Convention on the Law of the Sea (UNCLOS).	Defines "international rules and national legislation to prevent, reduce, and control pollution of the marine environment" (UNCLOS Part XII, Section 5).
London Protocol: The Convention on the Prevention of Marine Pollution by Dumping of Wastes and Other Matter 1972 (London Convention) and its 1996 Protocol (the London Protocol)	The London Convention and London Protocol are "international treaties of global application to protect the marine environment from pollution caused by the dumping of wastes and other matter into the ocean. In the United States, the Marine Protection, Research and Sanctuaries Act (MPRSA), also known as the Ocean Dumping Act, implements the requirements of the London Convention."[e]
MARPOL Annex V: Annex V of the International Convention for the Prevention of Pollution from Ships (MARPOL), 1973, as modified by the Protocol of 1978	2011: Resolution MEPC.201(62) Amendments to the Annex of the Protocol of 1978 Relating to the International Convention for the Prevention of Pollution from Ships, 1973 "Revised MARPOL Annex V" • Prohibits ship disposal of plastics into marine waters and imposes strict requirements for the disposal of other garbage.
Chemical- and Waste-Oriented Agreements Relevant to Plastic Waste and Pollution/Marine Debris	
Stockholm Convention: The Stockholm Convention on Persistent Organic Pollutants	"The Stockholm Convention on Persistent Organic Pollutants, adopted in 2001 and entered into force in 2004, is a global treaty whose purpose is to safeguard human health and the environment from highly harmful chemicals that persist in the environment and affect the well-being of humans as well as wildlife. The Convention requires parties to eliminate and/or reduce persistent organic pollutants (POPs), which have a potential of causing effects such as cancer and diminished intelligence and have the ability to travel over great distances."[f] • The United States is not yet a party to the Convention but does regulate some POPs. • A number of POP chemicals may be used as stabilizers in plastics or may be absorbed to plastic waste in the environment.[g] • "New chemicals can be added to the treaty based on a scientific review procedure."[h]

Basel Convention: The Basel Convention on the Control of Transboundary Movements of Hazardous Wastes and Their Disposal	2017: Thirteenth meeting of the Conference of the Parties to the Basel Convention–BC-13/11: Technical assistance; Work Programme 2018–2019. • By 2019, 187 countries added plastics to the Basel Convention. (BC-14/13 Fourteenth Meeting of the Conference of the Parties to the Basel Convention.) • Parties are required to control transboundary movements of the plastic waste covered under Convention procedures. All plastic waste and mixtures of plastic waste generated by Parties to the Convention which are to be moved to another Party are subject to the prior informed consent (PIC) procedure (the receiving party must agree in advance), unless they are non-hazardous and destined for recycling in an environmentally sound manner and almost free from contamination and other types of waste. The amendments themselves do not ban the import, transit, or export of plastic waste, but specify when and how the Convention applies to such waste. Technical guidance is in development (UNEP 2021). • The United States is *not* a party to the Basel Convention ◦ *New York Times* article on waste ban and the United States (March 2021).[i] ◦ U.S. Environmental Protection Agency guidance on applicability of Basel Convention to the United States.[j]

Biodiversity- and Species-Oriented Agreements Relevant to Plastic and Marine Debris

The Convention on Biological Diversity (CBD)	2010: Decision Adopted by the Conference of the Parties to the Convention on Biological Diversity at its Tenth Meeting (UNEP/CBD/COP/DEC/X/2) "The Strategic Plan for Biodiversity 2011–2020 and the Aichi Biodiversity Targets" 2016: CBD/COP/DEC/XIII/10 "Decision Adopted by the Conference of the Parties to the Convention on Biological Diversity–Addressing Impacts of Marine Debris and Anthropogenic Underwater Noise on Marine and Coastal Biodiversity" 2021: First Draft of the post-2020 Biodiversity Framework.[k] "Target 7. Reduce pollution from all sources to levels that are not harmful to biodiversity and ecosystem functions and human health, including by reducing nutrients lost to the environment by at least half, and pesticides by at least two thirds and eliminating the discharge of plastic waste."

continued

Global or Regional Organization	Legal Instrument and Relevant Coverage
Biodiversity- and Species-Oriented Agreements Relevant to Plastic and Marine Debris	
The Agreement for the Implementation of the Provisions of the United Nations Convention on the Law of the Sea of 10 December 1982 relating to the Conservation and Management of Straddling Fish Stocks and Highly Migratory Fish Stocks (United Nations Fish Stocks Agreement)	The Straddling Fish Stocks Agreement for implementing certain provisions of UNCLOS and Article 5(f) specifies that signatories "minimize pollution, waste, discards, catch by lost or abandoned gear, catch of non-target species, both fish and non-fish species, (hereafter referred to as non-target species) and impacts on associated or dependent species, in particular endangered species, through measures including, to the extent practicable, the development and use of selective, environmentally safe and cost effective fishing gear and techniques."
The Convention on the Conservation of Migratory Species of Wild Animals (CMS)	2014: UNEP/CMS/Resolution 11.30 Eleventh Meeting of the Conference of the Parties to the Convention on Migratory Species—Management of Marine Debris
G7 and G20 Frameworks and Charters	
Group of 20 (G20)	2017 *G20 action plan for marine debris* G20 Framework[l] • Website: Towards Osaka Blue Ocean Vision[m] • 2020 Report on Actions[n]
Group of 7 (G7)	2018 *G7 plastics charter.* Ocean Plastics Charter[o] (the United States is not a signatory) • 2019 Organisation for Economic Co-operation and Development Issue Brief: Improving Resource Efficiency to Combat Marine Plastic Litter[p] 2021 Carbis Bay G7 Summit Communiqué: Our Shared Agenda for Global Action to Build Back Better[q] (see para. 43, "We support...stepping up action to tackle increasing levels of plastic pollution in the ocean, including working through the UN Environment Assembly (UNEA) on options including strengthening existing instruments and a potential new agreement or other instrument to address marine plastic litter, including at UNEA-5.2.") • *2021 G7 Climate and Environment: Ministers' Communiqué,* London, May 21, 2021[r]

[a] See https://www.un.org/en/development/desa/population/migration/generalassembly/docs/globalcompact/A_RES_66_288.pdf.

[b] See https://wedocs.unep.org/handle/20.500.11822/10670.

[c] See http://pub.norden.org/temanord2020-535/temanord2020-535.pdf.

[d] See https://eia-international.org/ocean/plastic-pollution/legally-binding-agreement-on-plastic-pollution-faqs/.

[e] See https://www.epa.gov/ocean-dumping/ocean-dumping-international-treaties.

[f] See https://www.epa.gov/international-cooperation/persistent-organic-pollutants-global-issue-global-response#alaska?.

[g] See http://chm.pops.int/Implementation/PublicAwareness/PressReleases/POPRC16PressReleaseUV328elimination/tabid/8747/Default.aspx.

[h] See https://www.epa.gov/international-cooperation/persistent-organic-pollutants-global-issue-global-response#stockholm.

[i] See https://www.nytimes.com/2021/03/12/climate/plastics-waste-export-ban.html.

[j] See https://www.epa.gov/hwgenerators/new-international-requirements-export-and-import-plastic-recyclables-and-waste#fq4.

[k] See https://www.cbd.int/doc/c/abb5/591f/2e46096d3f0330b08ce87a45/wg2020-03-03-en.pdf.

[l] See https://sdg.iisd.org/news/g20-environment-ministers-adopt-framework-to-tackle-marine-litter/.

[m] See https://g20mpl.org/.

[n] See https://www.env.go.jp/press/files/en/872.pdf.

[o] See https://www.canada.ca/en/environment-climate-change/services/managing-reducing-waste/international-commitments/ocean-plastics-charter.html.

[p] See https://www.oecd.org/g20/summits/osaka/OECD-G20-Paper-Resource-Efficiency-and-Marine-Plastics.pdf.

[q] See http://www.g8.utoronto.ca/summit/2021cornwall/210613-communique.html.

[r] See https://www.gov.uk/government/publications/g7-climate-and-environment-ministers-meeting-may-2021-communique/g7-climate-and-environment-ministers-communique-london-21-may-2021.

REFERENCES

Karasik, R., T. Vegh, Z. Diana, J. Bering, J. Caldas, A. Pickle, D. Rittschof, and J. Virdin. 2020. *20 Years of Government Responses to the Global Plastic Pollution Problem: The Plastics Policy Inventory.* Durham, NC: Nicholas Institute for Environmental Policy Solutions, Duke University.

OECD (Organisation for Economic Co-operation and Development). 2018. "RE-CIRCLE: Resource Efficiency and Circular Economy Project." https://www.oecd.org/env/waste/recircle.htm.

UNEP (United Nations Environment Programme). 2018. Combating Marine Plastic Litter and Microplastics. https://www.gpmarinelitter.org/resources/information-documents/combating-marine-plastic-litter-and-microplastics-assessment.